NEURAL CREST
STEM CELLS
Breakthroughs and Applications

NEURAL CREST STEM CELLS
Breakthroughs and Applications

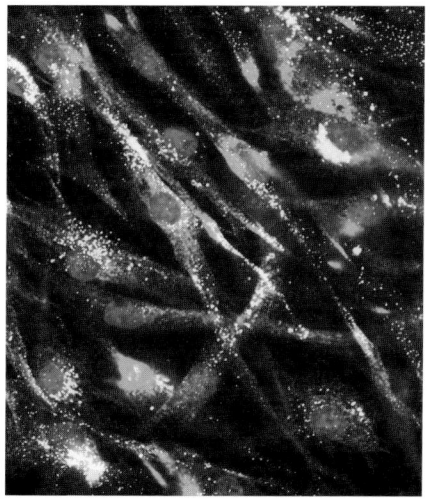

Editor

Maya Sieber-Blum
Newcastle University, UK

 **World Scientific**

NEW JERSEY · LONDON · SINGAPORE · BEIJING · SHANGHAI · HONG KONG · TAIPEI · CHENNAI

Published by

World Scientific Publishing Co. Pte. Ltd.

5 Toh Tuck Link, Singapore 596224

USA office: 27 Warren Street, Suite 401-402, Hackensack, NJ 07601

UK office: 57 Shelton Street, Covent Garden, London WC2H 9HE

British Library Cataloguing-in-Publication Data
A catalogue record for this book is available from the British Library.

NEURAL CREST STEM CELLS
Breakthroughs and Applications

ISBN-13 978-981-4343-80-0
ISBN-10 981-4343-80-3

Typeset by Stallion Press
Email: enquiries@stallionpress.com

Printed by FuIsland Offset Printing (S) Pte Ltd Singapore

CONTENTS

CONTRIBUTORS

**Sajjad Ahmad*
Royal Victoria Infirmary
Newcastle upon Tyne
United Kingdom

Department of Ophthalmology

Newcastle University
United Kingdom

Institute of Genetic Medicine
Newcastle University
Centre for Life
Newcastle upon Tyne
NE1 3BZ
United Kingdom

Haifa Ali
Institute of Genetic Medicine
Newcastle University
Centre for Life
Newcastle upon Tyne
NE1 3BZ
United Kingdom

* Corresponding author

Bill Chaudhry
Institute of Genetic Medicine
Newcastle University
Centre for Life
Newcastle upon Tyne
NE1 3BZ
United Kingdom

Oliver Clewes
Institute of Genetic Medicine and
North East England Stem Cell Institute
Newcastle University
Centre for Life
Newcastle upon Tyne
NE1 3BZ
United Kingdom

Simon J. Conway*
Developmental Biology and Neonatal Medicine Program
Herman B Wells Center for Pediatric Research
Indiana University School of Medicine
Indianapolis, IN 46202
USA
E-mail: siconway@iupui.edu

Deborah J. Henderson*
Institute of Genetic Medicine
Newcastle University
Centre for Life
Newcastle upon Tyne
NE1 3BZ
United Kingdom
E-mail: deborah.henderson@newcastle.ac.uk

---

* Corresponding author

Barbara Kaltschmidt*
Faculty of Biology
Universitätsstr. 25
Bielefeld University
D-33615 Bielefeld
Germany
E-mail: barbara.kaltschmidt@uni-bielefeld.de

Christian Kaltschmidt
Faculty of Biology
Universitätsstr. 25
Bielefeld University
D-33615 Bielefeld
Germany

Narihito Nagoshi*
Department of Physiology, Keio University
Department of Orthopedic Surgery, Keio University
School of Medicine, 35 Shinanomachi, Shinjuku-ku
Tokyo 160-8582
Clinical Research Centre, National Hospital Organization
Murayama Medical Centre
Tokyo 208-0011
Japan
E-mail: nagoshi@2002jukuin.keio.ac.jp

Hideyuki Okano
Department of Physiology
Keio University
School of Medicine
35 Shinanomachi, Shinjuku-ku
Tokyo 160-8582
Japan

---

* Corresponding author

Michael Olaopa
Developmental Biology and Neonatal Medicine Program
Herman B Wells Center for Pediatric Research
Indiana University School of Medicine
Indianapolis, IN 46202
USA

Charles Osei-Bempong
Institute of Genetic Medicine
Newcastle University
Centre for Life
Newcastle upon Tyne
NE1 3BZ
United Kingdom

Maya Sieber-Blum*
Institute of Genetic Medicine and
North East England Stem Cell Institute
Newcastle University
Centre for Life
Newcastle upon Tyne
NE1 3BZ
United Kingdom
E-mail: Maya.Sieber-Blum@newcastle.ac.uk

* Corresponding author

INTRODUCTION

In many respects the neural crest is a unique tissue. It is a transient embryonic tissue and a relatively recent acquisition during evolution, as it is only present in vertebrates. Neural crest cells arise in the neural folds of the forming neural tube, and while the neural tube gives rise to the spinal cord, neural crest cells undergo an epithelial-to-mesenchymal transformation, become migratory and translocate via different routes into the embryo where they generate a wide array of diverse cell types and tissues. Neural crest derivatives include the autonomic and enteric nervous systems, most primary sensory neurons, endocrine cells (adrenal medulla, calcitonin-producing cells of the thyroid), and pigment cells (melanocytes) of the skin and internal organs. Neural crest cells are instrumental in the formation of the cardiac outflow tract and they contribute to the smooth vasculature of the outflow tract and great vessels. Neural crest cells also give rise to the cranial mesenchyme, which generates craniofacial bone and cartilage, the corneal stroma, meninges, odontoblasts and striated musculature of the eye, among other tissues. Notably, while the origin of the neural crest is epidermal, its derivatives include both epidermal and mesenchymal cell types, underlining the versatility of this embryonic tissue.

It has long been recognised that migrating neural crest cells form a heterogeneous population of cells that consists of multipotent, oligopotent, bipotent and lineage-restricted stem cells. Moreover, neural crest stem cells have been identified in the periphery in various embryonic tissues, including the spinal and sympathetic ganglia, the epidermis, the cardiac outflow tract and the gastrointestinal system. More recently, evidence has emerged that neural crest-derived stem cells can persist into adulthood in various locations of the body.

While the mechanisms underlying neural crest cell lineage choice and differentiation have been investigated for decades, many are still elusive. Knowledge of these mechanisms is, however, essential for potential future clinical applications of neural crest-derived stem cells. For these reasons the first part of this book deals with some of the latest insights into the cellular and molecular mechanisms that drive neural crest stem cell differentiation.

The second part of the book focuses on neural crest-derived adult stem cells in various locations of the body. As many neural crest derivatives generate cell types of potential clinical importance, the presence and accessibility of these multipotent stem cells promises to open new avenues in cell based therapies and regenerative medicine. These aspects are explored in the second part of this book.

Maya Sieber-Blum
September 17, 2011

PART I

REGULATION OF EMBRYONIC NEURAL CREST CELL DIFFERENTIATION

1

THE NEW HEART FOR
THE NEW HEAD

Deborah J. Henderson and Bill Chaudhry

Institute of Genetic Medicine, Newcastle University, UK

Since the turn of the 20th century, experimental embryology and anatomy, in a host of species including human, have explained how the complex, three dimensional, twisting form of the fully septated heart develops from a simple tube. The simplicity of the accepted model — that the primary heart tube was already patterned in a recognisable order: Sinus venosus, atrium, right ventricle, left ventricle, outflow tract; with all the components already present at this early stage, was revised in 2001 when studies in chick and mouse demonstrated that cells are progressively added to both the arterial and venous poles, after the tube had already formed (Waldo *et al.*, 2001; Kelly *et al.*, 2001; Mjaatvedt *et al.*, 2001). This realisation has brought the vertebrate septated heart back into the evolutionary framework, as it emphasises the idea of a simple inflow chamber and an outflow chamber as the basic pattern upon which future modifications were made. The second change in thinking was arguably brought about by the emergence of stem cell biology. Cardiac stem cells were sought in primary and secondary heart fields, the epicardium, and endocardium. Against this backdrop, neural crest cells have moved back into the spotlight of cardiac development. Although, not truly pluripotent, — they cannot be differentiated into all of the lineages that are found in the embryo — they are multipotent with significant roles in forming and sculpting the modern vertebrate heart (Figure 1). In this review we will discuss the role of neural crest cells in general development and examine

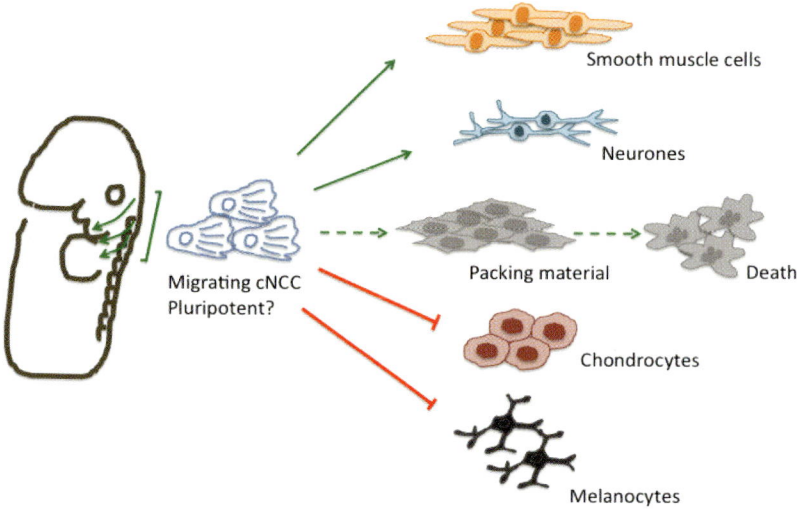

Figure 1. Cartoon showing the origin and fate of cardiac neural crest cells within the outflow tract of the heart. Cardiac neural crest cells originate in the hind brain region of the neural tube between the otocyst and somite 3 (green bracket). The neural crest cells migrate away from the neural tube, differentiating into smooth muscle cells and neurones within the outflow region. Many of the cardiac neural cells function transiently as packing material in the outflow cushions and in the aortic sac, then die by apoptosis once outflow septation is complete. Cardiac neural crest cells have the potential to differentiate into melanocytes and cartilage, although they are blocked from forming these lineages in the outflow region. Interestingly, neural crest cells do differentiate into melanoblasts in the atrioventricular valves.

the parallels that affect the heart. We will consider neural crest cells as honorary stem cells and ask which elements maintain them in their multipotent state and which processes cause them to progress through different lines of differentiation. Within this, we will consider the relevance of neural crest cells to human heart malformation.

New head — new heart

Neural crest cells play an important part in the evolutionary approach to developmental biology. In 1983, Gans and Northcutt argued that the emergence of neural crest cells allowed adaptations that enabled the organism to become a more efficient predator — the new head hypothesis. Their attractive hypothesis has been extended by bioinformatics analyses, and, with modification, the new head hypothesis is still respected. It is thought that the development of neural crest cells as a new tissue source may have evolved over a prolonged period, with previously developed gene pathways co-opted into new roles. Interestingly the other tissue that is involved in the formation of these "new head" structures is the mesoderm of the second heart field — which is contiguous with the facial mesoderm and has been shown to have common origins (Nathan *et al.*, 2008; Lescroart *et al.*, 2010). There is support for this from analyses of the sister group to the vertebrates, the tunicates, where progenitors that have properties of second heart field cells have been identified in the sea squirt *Ciona intestinalis* (Stolfi *et al.*, 2010). The neural crest cells that form the head are found adjacent to those that contribute to the cardiovascular system, with the latter arising between the otocyst (rhombomere 5) in the hind brain and somite 3 of the trunk. As such, they form at the boundary of cranial and trunk neural crest cells, and share properties of both (Scholl & Kirby, 2009).

Historically, a number of markers have been used to identify neural crest cells as they delaminate and migrate through the embryo. These markers have usually been non-specific, indicating migrating cells, for example HNK1 and alpha smooth muscle actin. It was the production of transgenic mice in which the Cre enzyme is driven by the Wnt1-Cre promoter (Danielian *et al.*, 1998), which is specific for the dorsal region of the neural tube and can be used, in conjunction with marker strains to permanently label neural crest cells and all their derivatives, that led to the definitive identification of mammalian neural crest cells as they migrate, and as they differentiate at the periphery (Jiang *et al.*, 2000). These analyses have confirmed many previous studies, demonstrating the huge contribution that the neural crest cells make to the developing and mature head (Figure 2). The evolution of the head is also linked to the development of a high pressure cardiovascular system. The most basic pattern of the heart has been considered to be a chamber, which, by cyclical reduction in its volume, expels fluid and enhances nutrient transfer. A single large heart in a closed circulatory system requires an antechamber that can accept the returning blood when the pumping chamber is contracted. This is the arrangement in animals with gills — for example the teleost zebrafish. Air breathing creates the need for two circulations, one to oxygenate blood

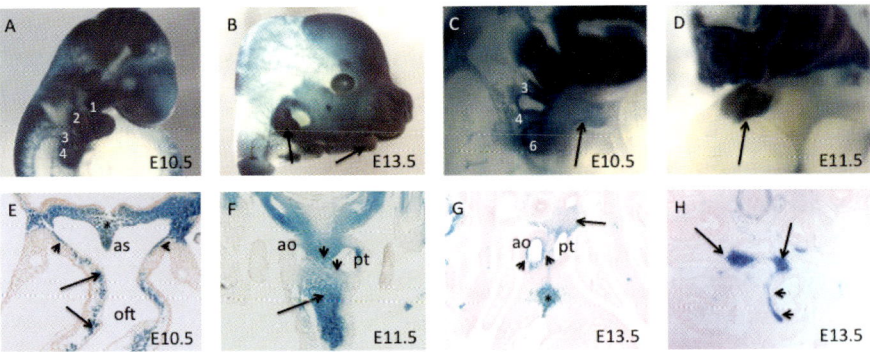

Figure 2. Neural crest cell distribution in the head and outflow regions. Neural crest cells are stained blue. (A, B) Neural crest cells make an extensive contribution to the vertebrate head, and can be found in the pharyngeal arches and their derivatives (arrows), in close proximity to the cardiac neural crest cells. (C–F) Neural crest cells migrate through the pharyngeal arch arteries (numbered 3–6 in C) and into the outflow tract of the heart (arrow in C) where they are abundant by E11.5, as outflow tract septation is initiated (arrow in D). Sections show neural crest cells in the lateral regions of the aortic sac (arrowheads) and outflow cushions (arrows) and in the dorsal wall of the aortic sac (asterisk) at E10.5 (E). A day later, outflow tract septation is initiated by the fusion of the dorsal wall of the aortic sac with the distal region of the outflow cushions (arrowheads in F show the fusion point). Neural crest cells are abundant throughout the outflow cushions by this time (arrow in F). (G, H) By E13.5, when septation is complete, neural crest cells line the lumen of the aorta (arrowheads in G) and are abundant in the ductus arteriosus (arrow) and the central region of the outflow septum (asterisk). The latter are the cells that are fated to die in the next gestational day. Cardiac neural crest cells also contribute to the cardiac plexus (arrows in H) and the parasympathetic innervation of the heart (arrowheads in H)

1–6 = pharyngeal arches; ao = aorta; as = aortic sac; oft = outflow tract; pt = pulmonary trunk.

across a high surface area and another to provide a pressurised flow throughout the body. To achieve this, we require formal valves and separation of flows into pulmonary and systemic streams. The second heart field precursors have already been linked with the evolution of the pulmonary circulation, as their lineage tracing indicates a major contribution to form the outflow tract and right ventricle, which is required for pumping blood into the lungs, and augmentation of the atria, to accept blood returning from the lungs (Moorman *et al.*, 2007; Perez-Pomares *et al.*, 2009). However, it is the neural crest that provides new valve and septal structures to make this work. Wnt1-cre lineage tracing again demonstrates that the cardiac structures that are made from neural crest cells are essential for air-breathing. The postnatal, fully septated, mammalian heart shows evidence of neural crest cells contributing to the valves, coronary arteries and the sympathetic innervation. The limited staining of the septated outflow (Figure 2) hides the huge contribution of neural crest cells during its development. Neural crest cell-derived smooth muscle cells within the pharyngeal region, surrounding the persisting pharyngeal arch arteries, are also obvious, but studies earlier in development have demonstrated the important active role they play in stabilising the aortic arch arteries (Nakamura, 1982; Bradshaw *et al.*, 2009). Thus, the emergence of neural crest cells, long ago in evolutionary history, was essential for the development of the vertebrate head, and the heart.

Journey to the poles of the heart

When the linear heart tube first forms (at E8.5 in the mouse heart, Carnegie stage 9 in the human), the outflow tract is the anterior part of the tube, linking the primitive ventricle with the symmetrical pharyngeal arch arteries. At this stage there is no separation of the chambers or outflow vessels apparent within the heart. It is composed of a myocardial layer and an endocardial layer separated by a thick extracellular matrix known as cardiac jelly. In some areas the cardiac jelly is more abundant and these areas are termed cushions. Over the next three days of mouse development, atria chambers become distinct from the ventricular chambers. The atria and ventricles separate right from left, and the initially single outflow vessel separates to form the aorta and pulmonary trunk. During the same time period, neural crest cells are being specified at the interface between neural plate and epidermis. The cardiac neural crest cells, arising in the posterior hind brain, stream through the paraxial mesoderm and into pharyngeal arches 3, 4, and 6. Some of these neural crest cells continue on to enter the heart late on E9.5. By this stage the heart tube, which was previously joined to the embryo by a mesentery along the length of its dorsal border, is attached to the mediastinum only at its arterial and venous poles. The first cardiac neural crest cells enter the abundant cardiac jelly of the outflow tract cushions at E9.5, with continual additions over the next 1–2 gestational days. The cardiac neural crest cells entering through the arterial pole of the heart pack into the cardiac jelly of the outflow cushions forming a dense compact array of cells. Remarkably, two days later almost all these neural crest cells are dead (Poelmann *et al.*, 1997). Neural crest cells enter through the venous pole approximately two days after those that had entered via the arterial pole (at E11.5) (Hildreth *et al.*, 2008). Not only is the neural crest cell contribution to the inflow region delayed in comparison with the outflow tract, but neural crest cells also enter the heart at the venous pole in much-reduced numbers. By E11.5, neural crest cell-derived nerves can be seen growing into the inflow of the heart, associated with the cardiac plexus, the main conduit for parasympathetic innervation of the heart (Figure 2; Hildreth *et al.*, 2008, 2009). Neural crest cells are also found in the region of the sinus node, suggesting that these cells provide the link between the pacemaker cells and the parasympathetic nervous system. In addition to this important role in cardiac innervation, limited numbers of neural crest cells are also found within the atrioventricular cushions, particularly associated with fusion seams between the inferior and superior cushions. Whether these cells play a significant role remains unclear, although it has been suggested they may differentiate into melanoblasts and play a role in stiffening the valve leaflets (Balani *et al.*, 2009). Other authors have suggested that they may play a role in fusion or sculpting of the atrioventricular cushions (Nakamura *et al.*, 2006; Hildreth *et al.*, 2009). There are a series of questions that immediately come to the fore. Are cardiac neural crest cells the same as other neural crest cells? What makes the cardiac crest cells migrate efficiently into the heart? Once there, what do they do and why do they die?

Are all neural crest cells the same?

An important question is whether the cardiac neural crest cells are the same as their cranial and trunk neighbours. In terms of initial specification and induction, neural crest cells are known to require a number of different signals, including bone morphogenetic proteins (BMPs), canonical Wnts, fibroblast growth factors (FGFs) and retinoic acid. This has been reviewed recently (Saukla–Spengler & Bronner–Fraser, 2008), and will not be discussed further here, except to say that all of these factors are also required for the induction and specification of the cardiac neural crest cell lineage. A feature that makes cardiac neural crest cells distinctly different from their neighbours is their origin from the neural tube between the fifth rhombomere and third somite. This area is patterned by the Hox code and two important signalling pathways — FGF and retinoic acid — are responsible for regulating Hox patterning in this area. Fgf8 is produced at the isthmus (midbrain-hindbrain boundary) and manipulation of Fgf8 expression at the isthmus has knock-on effects on pharyngeal arch patterning (Trainor *et al.*, 2002). Similarly, loss of Raldh2, the retinoic acid synthesising enzyme, severely disrupts hindbrain patterning, the Hox code, and the patterning of the pharyngeal arches (Niederreither *et al.*, 2000). It is also possible that sensitivity to hedgehog signalling may play a role in the behaviour of cardiac neural crest cells. It is known that if the dorsal part of the neural tube is ablated, cranial neural crest cells are able to regenerate but trunk and cardiac neural crest cells are not (Suzuki & Kirby, 1997). Recent studies in chicks have shown that this is related to hedgehog signalling. If hedgehog signalling is suppressed using cyclopamine in chicken embryos, cardiac neural crest cells regain the ability to regenerate following ablation of the dorsal region of the hindbrain neural tube (Hutson *et al.*, 2009). Moreover, the cardiac neural crest cells appear to function normally, migrating into the pharyngeal region from the regenerated neural folds and producing normal cardiac development. Thus, in this scenario, hedgehog signalling negatively regulates cardiac neural crest cell formation. Perhaps surprisingly, hedgehog signalling also plays, later, positive roles on cardiac neural crest cell ontogeny, as sonic hedgehog, secreted by the foregut endoderm, has also been shown to be required for survival of a subset of cardiac neural crest cells, and thus their colonisation of the outflow tract cushions, in mice (Goddeeris *et al.*, 2007).

One could argue that the pathway of migration of the cardiac neural crest cells is also a defining feature. This may be important, as in addition to their pre-migration patterning, cardiac neural crest cells are also exposed to differential signalling migrating through the pharyngeal arches. The mesenchyme and neural crest cells of arches 2 and 3 express different combinations of Eph receptors and ephrin ligands, which restrict the intermingling of neural crest cells within the different arches (Smith *et al.*, 1997; Robinson *et al.*, 1997; Mellott & Burke, 2008). Moreover neural crest cells respond to different combinations of extracellular matrix proteins, which can be either permissive (e.g. fibronectin) or non-permissive (e.g. versican), thus guiding the cells as they migrate to the periphery (Henderson & Copp, 1997; Perris & Perissinotto, 2000). Interactions of neural crest cells with the matrix, via enzymes such as matrix metalloproteinases (Cai & Brauer, 2002; Cantemir *et al.*, 2004) are essential for normal migration. Thus, as the neural crest cells migrate through the pharyngeal region and towards the outflow of the heart, they are exposed to a variety of extracellular (secreted) signalling molecules from a variety of sources. In turn, the neural crest cells themselves can alter the expression of specific signalling factors as they migrate. One important example is Fgf8. The cardiac neural crest cells are exposed to Fgf8 from a variety of sources during their travels; Fgf8 is expressed by the surface ectoderm and endoderm of the pharyngeal arches, as well as within the outflow tract of the heart itself. Reduction in Fgf8 levels within the pharyngeal region led to the death of cardiac neural crest cells within the pharyngeal arches despite initially normal specification and migration (Abu-Issa *et al.*, 2002). However, if there is a deficiency of neural crest cells, as demonstrated in neural crest cell-deficient chicken embryos, Fgf8 signalling is elevated in comparison with control embryos. The precise mechanism by which neural crest cell deficiency results in elevated Fgf8 signalling remains unclear. It may be that neural crest cells either breakdown or sequester the Fgf8, thus keeping levels within a margin thus allowing normal development. In either case, it is very interesting to note that this change in pharyngeal Fgf8 expression disturbs the addition of cells of second heart field origin to the outflow region of the heart (Hutson *et al.*, 2006). Moreover, it appears that excess Fgf8 can have detrimental effects on myocardial contractility in neural crest deficient embryos, which can ultimately result in heart failure (Farrell *et al.*, 2004; Hutson *et al.*, 2006). Thus, regulating cardiac neural crest cell numbers in the embryo may have implications beyond structural effects on outflow tract septation.

How do cardiac neural crest cells find their way?

The key guidance system required for migration of cardiac neural crest cells into the outflow region is the semaphorin-plexin-neuropilin axis.

Repulsive semaphorin ligands, 6A and 6B, are found in the pharyngeal region, lining the routes that will be taken by the cardiac neural crest cells. In contrast, neural crest cell-attracting semaphorin 3C ligand is expressed in the outflow tract myocardium. The sempahorin receptors, plexin A2, plexin D1, and neuropilin-1 are expressed upon the cardiac neural crest cells themselves. Thus, repulsive effects between semaphorin 6A/6B and plexin A2 in the pharyngeal region, in combination with the attractive effects of semaphorin 3C interacting with plexin D1 on the neural crest cell surface, guides them into the outflow tract (Brown *et al.*, 2001; Toyofuku *et al.*, 2008). Mouse mutants lacking semaphorin 3C develop outflow tract septation defects including common arterial trunk (Feiner *et al.*, 2001). Thus, this signalling pathway is essential for the appropriate migration of cardiac neural crest cells into the heart.

A second system that may also be relevant is provided by the platelet-derived growth factors *A* and *B* and their receptors α and β. Mutations in platelet derived growth factor receptor α, in both the Patch mouse and also in a gene targeted mutant, have been shown to result in outflow mal-formations including common arterial trunk and arch remodelling defects in a proportion of offspring, although cardiac neural crest cell survival appears grossly normal in these mutants. (Morrison-Graham *et al.*, 1992; Tallquist & Soriano, 2003). However, more recent studies have shown that neural crest cells carry both the α and β receptors and that the loss of both receptors in PDGFR α/β double mutants resulted in similar outflow defects, but with 100% penetrance (Richarte *et al.*, 2007). The ability of the neural crest cells to differentiate into smooth muscle cells within the pha-ryngeal arch arteries of the PRGFR α/β mutants appeared normal and no abnormalities in proliferation or cell death was observed. However, over the course of several days, progressively fewer neural crest cells were found in the outflow region, suggesting that these factors may be important for sustained migration of neural crest cells into the outflow tract (Richarte *et al.*, 2007).

Forming an orderly queue

Whilst the semaphorin guidance system is accepted as a mechanism for attracting neural crest cells into the outflow region, there is still a controversy regarding the cell autonomous mechanisms that the neural crest cells use to get to the outflow tract. Recent studies in Xenopus and zebrafish embryos have suggested that contact inhibition of locomotion mediates neural crest cell migration both *in vitro* and *in vivo*, and that this is regulated by planar cell polarity signalling, via RhoA and ROCK (Carmona-Fontaine *et al.*, 2008). Thus, it is suggested that transient contact between travelling, adjacent, neural crest cells causes them to move apart, and that the space created by this cellular repulsion allows other cells to be released from contact inhibition. In this model, contact between adjacent neural crest cells results in their separation, meaning that neural crest cells migrate as isolated individuals, but part of an interacting group. Neural crest cells in the chick have also been studied, but in this system the neural crest cells migrate as a continuous stream, with the cells maintaining continuous direct communication with their neighbours (Teddy & Kulesa, 2004; Kulesa *et al.*, 2008). It is possible that there are slightly different mechanisms that produce the same effect between the chick and the other vertebrates. Data from fixed mouse embryos and genetic models suggests that communication between neural crest is important, but does not indicate whether either the chick or frog models will be dominant. Live imaging of mouse embryos is technically difficult but will yield interesting results.

Whether neural crest cells invite or repulse inter-cellular contact, the mechanism will require an inter-cellular communication event. Gap and adherens junctions are two possible routes. Gap junctions are ionic connections between cells. They are formed from connexins, with connexin 43 implicated in cellular communication and the regulation of polarised cell movements within cardiac neural crest cells (Xu *et al.*, 2006). Interestingly, there appears to be interactions between the gap junction system and the cytoskeleton as both focal adhesions and actin filament architecture were disturbed in connexin 43-deficient as well as connexin 43 over-expressing cardiac neural crest cells. Disturbing connexin 43 in cardiac neural crest cells altered the response of the cells to semaphorin 3C (Xu *et al.*, 2006), one of the molecules important in the chemo-attraction of neural crest cells into the outflow tract (see earlier), showing that links exist between the different levels of migratory regulation of neural crest cells. Although connexin 43 knockout mice die at birth with outflow tract obstruction and abnormalities in the right ventricle (Reaume *et al.*, 1995; Huang *et al.*, 1998), specific deletion of connexin 43 from neural crest cells did not result in structural heart defects, even though the coronary artery patterning was disturbed (Kretz *et al.*, 2006; Liu *et al.*, 2006). In contrast, deletion of connexin 43 from the entirety of the neural tube recapitulated the outflow abnormalities observed in the knockout mice, with excessive numbers of neuroepithelium-derived cells found in the outflow region. Thus, connexin 43 may be modulating the transformation, delamination, or migration of neural crest cells towards the heart, with indirect effects on the development of the outflow region (Liu *et al.*, 2006) or there may be functional redundancy relating to neural crest cell communications.

N-cadherin is not expressed at high levels in neural crest cells as they migrate, but is up regulated as the cells condense in the outflow cushions (Luo *et al.*, 2006). When N-cadherin was conditionally deleted from neural crest cells, these cells migrated to the outflow tract efficiently, but remained rounded, failed to align and lost contact with the neighbouring cells, resulting in outflow tract septation defects, predominantly in the common arterial trunk (Luo *et al.*, 2006). This suggests that the adherens junction is important in intercellular communication between neural crest cells within the outflow region, rather than during their migration. In keeping with this local effect, outflow tract rotation was incomplete in the N-cadherin and Wnt1cre embryos, suggesting that N-cadherin might regulate a process required for normal rotation of the outflow cushions, presumably via its effects on neural crest cell adhesion or polarisation (Luo *et al.*, 2006). Similar findings were observed when focal adhesion kinase (FAK), a mediator of signal transduction by integrins and growth factor receptors, was deleted specifically in neural crest cells (Vallejo–Illarramendi *et al.*, 2009). Although normal long-range migration of cardiac neural crest cells was observed, outflow tract malformations including common arterial trunk and double outlet right ventricle were present. The cardiac neural crest did not appear to condense properly within the outflow tract cushions in the mutant embryos, suggesting that some behaviour patterns of the neural crest cells were compromised. This was supported by the finding that the actin cytoskeleton was disrupted in the FAK-deficient neural crest cells entering the outflow, resulting in a more rounded morphology than was observed in control embryos.

However, the neural crest cells were found to differentiate into smooth muscle cells in the outflow tract, suggesting that the differentiation was not prevented. These findings in FAK and N-cadherin mutants suggest that elongation and alignment of the neural crest cells within the outflow region is required for normal outflow septation. Understanding these cellular changes, and the effects that they have on the architecture of the septating tissues will be important for understanding how outflow tract septation and remodelling is achieved.

What do the neural crest cells do in the cardiac outflow?

Neural crest cells are multipotent cells, capable of differentiating into a wide range of cell types, including smooth muscle cells, neurones, cartilage and melanocytes, to name a few. However, within the confined regions of the outflow tract, the differentiation of these multipotent cells appears to be severely restricted. During normal development, the cardiac neural crest cells within the outflow tract differentiate into smooth muscle cells in the aorta and pulmonary trunk arterial walls, contribute to the walls of the proximal coronary vessels, make a permanent contribution to the semilunar valve leaflets, and give rise to the neurones that form the sympathetic innervation of the arterial pole of the heart (Jiang *et al.*, 2000). A small number also differentiate into melanocytes, particularly in the atrioventricular valves (Brito & Kos, 2008). Cardiac neural crest cells first enter the outflow tract at E9.5 of mouse development, although they are found within the adjacent pharyngeal arches prior to this. Cardiac neural crest cells enter the outflow tract laterally, where they contribute to the outflow tract cushions, and also medially, via the dorsal wall of the aortic sac (Figure 2). At E10.5, two outflow cushions can be seen positioned rostrally and caudally within the outflow vessel along its entire length. Initially there are only limited numbers of neural crest cells in the cushions, but by E11.5, when outflow tract septation is initiated, the neural crest cells pack the distal outflow cushions as well as the dorsal wall of the aortic sac (Figure 2). This rapid increase in neural crest cell numbers is achieved by continued migration of neural crest cells into the outflow cushions, and also by rapid proliferation of those neural crest cells already within the cushions (Chen *et al.*, 2007). Outflow tract septation is initiated when the remodelling of the pharyngeal and outflow regions, together with the expansion of these structures due to increased neural crest cell numbers, brings the dorsal wall of the aortic sac, between the origins of the fourth and sixth pharyngeal arch arteries, into contact with the most distal regions of the expanded outflow cushions (Anderson *et al.*, 2010). Fusion of this region of the dorsal wall of the aortic sac with the two outflow cushions, simultaneous with the fusion of the two cushions with each other, initiates outflow septation and separates the systemic and pulmonary circulations. Thus, although both the dorsal wall of the aortic sac and the outflow cushions are covered by a layer of endocardium, and fusion occurs between these two endocardial layers, the bulk of fusing structures are derived from neural crest cells. Once initiated, outflow septation proceeds rapidly, progressing in a distal to proximal wave along the outflow vessel, and completed two days later (in mouse) by the fusion of the most proximal regions of the outflow cushions with the atrioventricular cushions and with the ventricular septum. This fusion process walls the aorta into the left ventricle and the pulmonary trunk into the right ventricle, finally separating the systemic and pulmonary circulations.

Lineage tracing of crest cells as they form smooth muscle cells in the outflow has shown that they initially contribute only to the adjacent or facing walls of the outflow vessels. This is because septation of the outflow requires the fusion of the dorsal wall of the aortic sac with the longitudinal outflow cushions which are filled with neural crest cells. The outer part of the forming aorta and pulmonary artery are devoid of neural crest-derived smooth muscle at this point. However, the neural crest-derived smooth muscle eventually forms a complete layer that surrounds the endothelium of the vessels (Figure 2). In a similar fashion, the neural crest contributes to the posterior right and left leaflets of the pulmonary valve and the anterior right and left coronary leaflets of the aortic valve, as these originate from the neural crest cell-packed outflow cushions that have septated the outflow tract. The remaining valve leaflets which are formed from the lateral cushions, are composed of cells derived from the endocardium, and thus have a much-reduced contribution from the neural crest. Recent studies have shown that the neural crest cells that contribute to the outflow tract cushions are essential for the functional maturation of the valves (Jain *et al.*, 2011). Both FGF and BMP signalling has been implicated in promoting the differentiation of neural crest cells within the outflow cushions, and thus, to indirectly regulate valve development (Zhang *et al.*, 2010).

Fated to die?

Neural crest cells are a largely transient population in the outflow of the heart. Large numbers migrate into the outflow, creating the bulk of the septating structures in the outflow vessel and playing a major role in the septation process itself, but then almost completely disappear once outflow septation is complete (Poelmann *et al.*, 1998; Jiang *et al.*, 2000). Remarkably, extensive apoptosis removes the neural crest cells from the outflow tract septum, peaking in the period immediately following the completion of outflow tract septation (Poelmann *et al.*, 1998). The reason for this remains unclear. It has been suggested that the neural crest cells release inductive factors for the muscularisation of the septum itself, by the process of myocardialisation (Poelmann *et al.*, 1998). However, this process is well underway before there is any significant cell death within the outflow cushions (van den Hoff *et al.*, 1999) and it seems unlikely that apoptosing cells would release signalling molecules as the surrounding cells engulf them. Although less attractive as a theory, it may be that the major role of neural crest cells in the outflow tract, at least during the process of septation, is simply to provide physical bulk (packing material) to the septating structures, the outflow cushions and the dorsal wall of the aortic sac. This would be a method of positioning the cushions in apposition against the pulsatile hydrostatic forces of blood pumping out through the outflow. Concomitant with the septation of the outflow vessel, the proximal region of the forming outlet septum becomes muscularised, as the myocardial cells within the outflow tract wall move into the outflow cushions (van den Hoff *et al.*, 1999; Phillips *et al.*, 2005). Over a period of four days, in the mouse, from E11.5–E14.5, the forming proximal outlet septum becomes completely muscularised. Although it remains unclear what induces the myocardial cells within the outflow tract wall to change phenotype and move into the outflow cushions, the timing of neural crest cell colonisation of the cushions makes it possible to involve them in inducing this process. Indeed, the final muscularisation of the central region of the septum is only achieved as the last of the neural crest cells die in this region at E14.5, suggesting that the death of neural crest tissue may be required to make space for the myocardial cells within the septum, or even may be inducing their migration.

Stabilising the pharyngeal arch arteries

Neural crest cells have been known to play an important role in the maintenance of the pharyngeal arch arteries for many years. Neural crest cells contribute the bulk of the mesenchyme within the pharyngeal arches and surround the pharyngeal arch arteries from their first appearance. Recent studies have suggested that although endothelial vessels can form in neural crest cell-deficient pharyngeal arches, these vessels rapidly regress and disappear (Bergwerff *et al.*, 1998; Bradshaw *et al.*, 2009). Thus, neural crest cells appear to be essential for the stabilisation of the pharyngeal arch arteries rather than their formation, supporting and maintaining the nascent vessels even before smooth muscle cell differentiation occurs. Later during the development, neural crest cells produce much of the smooth muscle layer that surrounds the maturing vessels. This varies considerably in different arteries. For example, the ductus arteriosus, which is derived from the left sixth aortic arch artery is a largely muscular artery, with only minimal elastic components. The smooth muscle cells in this transient vessel are almost entirely derived from neural crest cells (Figure 2). In contrast, the aortic arch is an elastic artery, having multiple layers of elastic fibres distributed within its muscular intima (Bergwerff *et al.*, 1999).

Disruption of a number of secreted factors is associated with abnormal remodelling of the pharyngeal arch arteries. For example, disruption of endothelin-1 signalling in the pharyngeal region results in apparently normal formation of the pharyngeal arches, but abnormal regression of pharyngeal arches 4 and 6 as development proceeds, suggesting a role for this signalling pathway in arch remodelling (Kurihara *et al.*, 1995; Yanagisawa *et al.*, 1998). The endothelin-A receptor is expressed by neural crest cells themselves, whereas endothelin-1, and the enzyme which is required for cleavage to its active form, endothelin-converting enzyme (ECE1) is expressed in the epithelia of the pharyngeal arches (Kempf *et al.*, 1998; Clouthier *et al.*, 1998; Clouthier *et al.*, 2003). Interestingly, smooth muscle cell differentiation (as shown by SM22a-lacZ expression) was impaired in ECE1-/- and ETA-/- mutants when compared with stage-matched controls (Yanagisawa *et al.*, 1998), suggesting that this may be a primary factor in the failure to maintain the arch arteries.

Neural crest survival and differentiation

Cardiac neural crest cells are required to proliferate during their migration from the neural tube to the heart, but to differentiate when they reach their targets in the pharyngeal arches and the outflow tract. It may be that neural crest cells are a population that are programmed to die unless they are actively maintained — perhaps as a safeguard against ectopic seeding and teratoma formation. A number of factors have been implicated in the proliferation or survival of neural crest cells in the pharyngeal region. Sonic hedgehog, secreted by the endoderm as discussed earlier, is one such factor (Goddeeris *et al.*, 2007). Transforming growth factor (TGFβ) signalling has also been implicated in neural crest cell survival, particularly within the pharyngeal and outflow regions. Both TGFβ and BMP ligands signal via a complex of type I and type II TGFBR, which differ according to the precise TGFβ or BMP ligand involved, and the cellular context (reviewed in ten Dijke and Arthur, 2007). Specific disruption of TGFBR1 (Alk5), in neural crest cells, results in common arterial trunk and aortic arch remodelling defects (Wang *et al.*, 2006). In these mutants the neural crest cells appear to migrate normally, but there is extensive cell death in the pharyngeal region, prior to and during the period when outflow tract septation would normally take place. These data suggest a role for TGFBR1 in cardiac neural crest cell survival within the outflow region. Disruption of the BMP type I receptor, Alk2, in neural crest cells also results in common arterial trunk and pharyngeal arch defects. The pharyngeal arches appear to form normally in this mutant, but show reduced smooth muscle cell differentiation around them, suggesting there is insufficient neural crest cells to stabilise the nascent vessels (Kaartinen *et al.*, 2004). It is likely the common arterial trunk results from impaired migration of neural crest cells into the outflow tract. Mouse mutants for BMPR1A (Alk6) support this as they have reduced numbers of crest cells in the outflow cushions suggesting that BMPR1A may be required for proper colonisation of the cushions and formation of a septated outflow tract (Stottmann *et al.*, 2004). Interestingly, the BMPR1A mutants have a shortened outflow tract and reduced right ventricle followed by heart failure at mid gestation. This led the authors to speculate that BMPR1A is required in neural crest cells for production of a downstream signal that induces movement of second heart field cells to form the outflow tract and right ventricle (Stottman *et al.*, 2004).

Conditional disruption of TGFBR2 in neural crest cells also results in common arterial trunk and pharyngeal arch patterning defects (Choudhary *et al.*, 2006), but in this case, there was no evidence of increased apoptosis in the neural crest cells. In fact, outflow tract septation failed in the context of apparently normal levels of neural crest cells within the pharyngeal and outflow region and their normal differentiation into smooth muscle cells. The cause of the failure in outflow tract septation in this case remains unclear, although it would seem likely that there is disruption of some cellular characteristic that is essential for the initiation of outflow tract septation. Choudhary *et al.* (2006) speculate that the neural crest cells undergo premature differentiation in the distal outflow tract when TGFBR2 is deleted in neural crest cells, although the phenotype of these cells remains unclear. Loss of TGFBR2 from neural crest cells also results in an interrupted aortic arch, although this was associated with a marked increase in dying cells in the regressing arch artery. Thus, it appears that the TGFBR2 must be playing distinct roles in the pharyngeal arch arteries and within the distal outflow tract.

Although several factors have been implicated in differentiation of neural crest cells into smooth muscle cells within the outflow region, surprisingly little is known about what induces cardiac neural crest cells to form neurons within the outflow region, or what regulates their extensive death following septation. As cardiac neural crest cells are known to have broad differentiation potential, being capable *in vitro* at least of differentiating into a wider range of cell types than are normally found in the normal heart, including melanocytes and cartilage (Ito & Sieber–Blum, 1991; Sieber–Blum & Ito, 1995), an equally pertinent question is what restricts their differentiation to specific lineages within the environment of the outflow tract. Mice lacking c-Kit, which is known to be important in skin melanisation do not have melanocytes within their hearts, suggesting similar mechanisms may regulate melanocyte differentiation in these distinct organs. In contrast, mice over-expressing endothelin-3 have excessive numbers (Britto & Kos, 2008). Thus, mechanisms exist to restrict melanocyte differentiation from neural crest cells to the atrioventricular valves; strikingly, very few are observed in the outflow region. A clue about how neural crest cells may be restricted from forming cartilage came from a recent paper in which the transcription factor Ets1 was functionally inactivated in mouse embryos (Gao *et al.*, 2010). In Ets1 null embryos, abnormal nodules of cartilage were found in the peri-aortic region, which were shown to be derived from neural crest cells. Ets1 acts downstream in a signalling pathway regulated by FGF, via the kinases MEK and ERK that inhibit cartilage formation (Bobick & Kulyk, 2004). Chemical inhibition of MEK in mouse hearts in culture resulted in cartilage formation, thus suggesting that this pathway might restrict differentiation of cardiac neural crest cells within the mammalian heart (Gao *et al.*, 2010). Interestingly, cartilage nodules have also been found in the hearts of mouse embryos expressing a dominant-negative form of connexin 43 (Sullivan *et al.*, 1998). Thus, alterations in cellular behaviour effected by disrupting neural crest cell communication, or by direct effects of connexin 43 on MEK/ERK signaling (Plotkin *et al.*, 2002; Stains & Civitelli, 2005), might result in the abnormal differentiation of cardiac neural crest cells in the hearts of these mice.

What heart malformations can be attributed to neural crest cells?

The first indication that neural crest cells might be associated with congenital heart defects emerged when it was shown that ablation of the neural folds, between the otocyst and third somite, resulted in defects in outflow tract septation (Kirby *et al.*, 1983). The principal defect arising from this ablation was common arterial trunk (persistent truncus arteriosus), in which the outflow tract remains completely unseptated, leaving a single outflow vessel connected to the right ventricle. However, other malformations, including double outlet right ventricle (where the aorta and pulmonary trunk both take their origin from the right ventricle) and transposition of the great arteries (where the aorta and pulmonary trunk are transposed, such that the aorta exits from the right ventricle whereas

the pulmonary trunk exits from the left ventricle) were found in a proportion of the ablated embryos (Kirby *et al.*, 1983; reviewed in Hutson &
Kirby, 2007). The ablation of the neural folds in chick embryos remained
the favoured model for the study of cardiac neural crest cells until the
advent of genetic mouse models in the late 1980s and early 1990s. One of
the most important mouse models has been the *Splotch* mouse (Franz,
1989; Conway *et al.*, 1997 a, b; Epstein *et al.*, 2000; Bradshaw *et al.*, 2009).
Splotch is an allelic series of mouse mutants, carrying mutations in the
Pax3 gene (Epstein *et al.*, 1991), a homeobox-containing transcription factor that is required for neural crest cell expansion within the dorsal part of
the neural tube (Conway *et al.*, 2000). Mutations in Pax3, as seen in *Sp*
and *Sp^{2H}* mutants, are associated with abnormalities in outflow tract septation that are similar to those seen in chick ablation models. These
include common arterial trunk, double outlet right ventricle, ventricular
septal defects and pharyngeal arch remodelling defects (Franz, 1989).
Recent studies have shown that *Sp^{2H}* embryos have a marked deficiency of
neural crest cells, resulting in hypoplasia of both the outflow cushions and
the dorsal wall of the aortic sac (Epstein *et al.*, 2000; Bradshaw *et al.*, 2009).
The reduced size of these structures means that they fail to become
apposed to initiate outflow septation, resulting in common arterial trunk.
However, analysis of this mutant has revealed an unexpected aspect of
outflow tract development. It appears that it is the maintenance of the
sixth pharyngeal arch arteries, rather than the extent of cushion hypoplasia, which determines whether the embryos develop common arterial
trunk or double outlet right ventricle. In cases where the sixth arch arteries
were missing, a common arterial trunk was found, whereas when these
were retained, double outlet right ventricle was manifest (Bradshaw *et al.*,
2009). This initially surprising finding can be explained by the central role
for the dorsal wall of the aortic sac in the septation process. In cases where
the sixth pharyngeal arches are missing, the region of the dorsal wall of the
aortic sac which fuses with the outflow cushions, which lies between the
origins of the fourth and sixth pharyngeal arch arteries, is misplaced or
markedly deficient. As a consequence, the dorsal wall of the aortic sac does
not become apposed to the outflow cushions, and outflow tract septation
fails (Figure 3). In contrast, when the sixth aortic arteries are present, the
dorsal wall of the aortic sac between the origins of the fourth and sixth

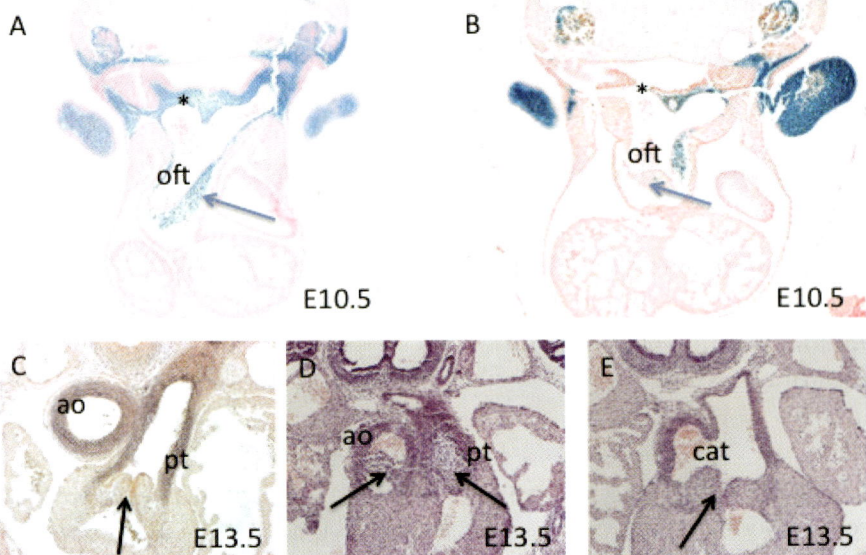

Figure 3. Outflow tract septaton defects result from abnormalities in neural crest cells in the outflow tract. Neural crest cells are stained blue. (A, B) Neural crest cells are deficient both in the outflow tract cushions (arrows) and in the dorsal wall of the aortic sac (asterisk) at E10.5 in the Sp^{2H}/Sp^{2H} (B) embryos compared to wild type controls (A). (C–E) The aorta and pulmonary trunk are separated and spiral around one another in the normal heart, with the valves offset (arrows in C). In contrast, in neural crest-deficient hearts with double outlet right ventricle (D) the aorta and pulmonary trunks are parallel, with the valves at the same level (arrows). In more severely affected neural crest-deficient hearts, with common arterial trunk (E), the aorta and pulmonary trunk are in continuity and the valve leaflets are shared (arrows).

ao = aorta; cat = common arterial trunk; oft = outflow tract; pt = pulmonary trunk.

pharyngeal arch arteries, although hypoplastic, achieves apposition with the outflow cushions, and thus septation is initiated. The abnormal spiralling (indeed the failure of spiralling) of the cushions in this latter scenario means that the aorta forms rightward of the developing pulmonary trunk, and thus cannot achieve continuity with the left ventricle (Bradshaw *et al.*, 2009). Thus, the mouse models of neural crest deficiency suggest a spectrum of defects from common arterial trunk at the severe end of the spectrum, to peri-membranous ventricular septal defects at the mild end, with defects such as double outlet right ventricle and over-riding aorta in between. The continued analysis of genetic models of neural crest cell deficiency, where a variety of genes are deleted in neural crest cells, is likely to further elucidate how specific cardiac malformations arise.

Cardiac neural crest cells and human malformation

The neural crest is thus a crucial component of normal heart development. Once identified as such, relationships with human congenital heart malformation have been sought. Despite initial optimism (Siebert *et al.,* 1985), this has not been a fruitful search. Two well-known syndromes with abnormalities in structures derived from neural crest are DiGeorge syndrome and CHARGE syndrome. Although not directly caused by abnormal neural crest contribution, these syndromes have begun to indicate the importance of interactions of neural crest cells with pharyngeal artery and second heart field precursors.

One of the first human genetic syndromes to be associated with neural crest deficiency was DiGeorge syndrome (velocardiofacial syndrome; CATCH22). This syndrome typically results from the deletion of a 11Mb region of the long arm of chromosome 22 and is associated with a wide range of developmental defects, of which outflow tract and pharyngeal arch abnormalities play a significant part (OMIM: 188400). The types of malformations occurring in combination with these heart defects, which include craniofacial abnormalities and hypoplasia of the thymus and parathyroid glands, led to the consideration of DiGeorge syndrome as a neurocristopathy resulting from a deficiency of neural crest cells during embryonic development. However, a variety of studies over the past 10 years have suggested that the cardiovascular defects observed in DiGeorge syndrome, at least, may not arise from a primary abnormality in neural crest cells. One of the genes in the DiGeorge critical region, TBX1, has been identified as the gene most likely to be responsible for the cardiac malformations in DiGeorge syndrome. In the mice, haploinsufficiency for Tbx1 results in defects, including an interrupted aortic arch, that closely resemble the cardiovascular malformations in human patients (Merscher *et al.*, 2001; Lindsay *et al.*, 2001; Jerome & Papaioannou, 2001). Complete loss of Tbx1 function is associated with common arterial trunk, a malformation found in a subset of the DiGeorge Syndrome patients. However,

Tbx1 is not expressed in neural crest cells, but instead is found localised to the ectoderm and mesoderm of the pharyngeal region, overlapping with the second heart field. Cells originating in the second heart field make major contributions to the outflow (and inflow) regions of the heart tube after its initial formation, and loss of these cells results in phenotypes that overlap with those resulting in neural crest cells deficiency, including common arterial trunk and double outlet right ventricle (Waldo *et al.*, 2001). Importantly, the neural crest cells within the pharyngeal region are found in close proximity to the second heart field cells where it has been shown that there is reciprocal signalling between the two cell types. Analysis of neural crest cells distribution in Tbx1 null mice has shown that neural crest cells migration to the heart is disturbed (Vitelli *et al.*, 2002), showing how disruption of one's cell type can secondarily affect neighbouring cell types. This may go some way to explaining the resemblance of DiGeorge syndrome to a neurocristopathy, although the direct requirement for cells of second heart field origin in both cardiac and craniofacial development may account for its initial mislabelling.

CHARGE syndrome (OMIM: 214800) is an inherited syndrome that in many respects resembles DiGeorge syndrome. Patients with CHARGE syndrome have a range of defects that include choanal atresia, coloboma of the eye and abnormalities of the inner ear, and in many cases outflow malformations, particularly interrupted aortic arch. CHARGE syndrome is typically caused by mutations of the chromatin remodelling factor CHD7, which binds to the regulatory regions of thousands of developmental genes (Schnetz *et al.*, 2010). In contrast to Tbx1, however, Chd7 is expressed in neural crest cells and CHARGE syndrome thus may be a primary neurocristopathy (Sanlaville *et al.*, 2006). Studies in frogs have shown that ChD7 is required for the migration of neural crest cells, and that its loss prevents their migration to the periphery (Bajpai *et al.*, 2010). Moreover, knocking down Chd7 expression in frog embryos resulted in abnormal positioning of the outflow vessel, which the authors liken to the cardiac malformations seen in CHARGE patients (Bajpai *et al.*, 2010). However, another study carried out in mouse embryos, suggests that it is the expression of Chd7 in the surface ectoderm of the pharyngeal arches, rather than neural crest cells that is important for normal patterning of the pharyngeal arch arteries (Randall *et al.*, 2009). Thus, Chd7 might be required in multiple tissues to bring about normal development of the cardiac outflow. The question does remain as to why human patients with congenital heart disease do not commonly present mutations in genes that are important in neural crest cells, at least so far. Mutations in the Pax3 gene are found in patients with Waardenberg syndrome (OMIM: 193500). Patients with this syndrome present a combination of pigmentation abnormalities and hearing defects, both of which are associated with neural crest cell deficiency, but they do not have cardiac defects. Although Waardenburg patients carry a single mutated copy of Pax3, and cardiac defects are only seen in mouse embryos with two mutated copies of the mouse Pax3 gene, it is also possible that neural crest abnormalities in human development are embryonically lethal, perhaps because of the accompanying myocardial dysfunction that has been seen in mouse embryos (Farrell *et al.*, 2004; Hutson *et al.*, 2006).

Dramatic advances in our understanding of how neural crest cells contribute to heart's development have been made over the past 30 years. These have shown that neural crest cells play multiple roles in cardiac development and also influence other cell types within the outflow region. It remains to be established whether neural crest cell deficiency, or more subtle defects in neural crest fate or function, make a significant contribution to human cardiac malformation. However, the identification of neural crest stem cells within the mouse heart (Tomita *et al.*, 2005) and the recent discovery that cardiomyocytes, derived from neural-crest cell stem cells are found in the peri-infarct region following myocardial infarction (Tamura *et al.*, 2011), suggests that neural crest cells may have even more to offer to the heart, than has previously been imagined.

References

Abu-Issa R, Smyth G, Smoak I, Yamamura K, Meyers EN (2002). Fgf8 is required for pharyngeal arch and cardiovascular development in the mouse. *Development* 129(19): 4613–4625.

Anderson RH, Cook A, Brown NA, Henderson DJ, Chaudhry B, Mohun T (2010). Development of the outflow tracts with reference to aortopulmonary windows and aortoventricular tunnels. *Cardiol. Young* 20(Suppl 3): 92–99.

Bajpai R, Chen DA, Rada-Iglesias A, Zhang J, Xiong Y, Helms J, Chang CP, Zhao Y, Swigut T, Wysocka J (2010). CHD7 cooperates with PBAF to control multipotent neural crest formation. *Nature* 463(7283): 958–962.

Balani K, Brito FC, Kos L, Agarwal A (2009). Melanocyte pigmentation stiffens murine cardiac tricuspid valve leaflet. *J. R Soc. Interface* 6(40): 1097–1102.

Bergwerff M, Verberne ME, DeRuiter MC, Poelmann RE, Gittenberger-de Groot AC (1998). Neural crest cell contribution to the developing circulatory system: Implications for vascular morphology? *Circ. Res.* 82(2): 221–231.

Bergwerff M, DeRuiter MC, Hall S, Poelmann RE, Gittenberger-de Groot AC (1999). Unique vascular morphology of the fourth aortic arches: Possible implications for pathogenesis of type-B aortic arch interruption and anomalous right subclavian artery. *Cardiovasc. Res.* 44(1): 185–196.

Bobick BE, Kulyk WM (2004). The MEK-ERK signaling pathway is a negative regulator of cartilage-specific gene expression in embryonic limb mesenchyme. *J. Biol. Chem.* 279(6): 4588–4595.

Bradshaw L, Chaudhry B, Hildreth V, Webb S, Henderson DJ (2009). Dual role for neural crest cells during outflow tract septation in the neural crest-deficient mutant Splotch(2H). *J. Anat.* 214(2): 245–257.

Brito FC, Kos L (2008). Timeline and distribution of melanocyte precursors in the mouse heart. *Pigment Cell Melanoma Res.* 21(4): 464–470.

Brown CB, Feiner L, Lu MM, Li J, Ma X, Webber AL, Jia L, Raper JA, Epstein JA (2001). PlexinA2 and semaphorin signaling during cardiac neural crest development. *Development* 128(16): 3071–3080.

Cai DH, Brauer PR (2002). Synthetic matrix metalloproteinase inhibitor decreases early cardiac neural crest migration in chicken embryos. *Dev. Dyn.* 224(4): 441–449.

Cantemir V, Cai DH, Reedy MV, Brauer PR (2004). Tissue inhibitor of metalloproteinase-2 (TIMP-2) expression during cardiac neural crest cell migration and its role in proMMP-2 activation. *Dev. Dyn.* 231(4): 709–719.

Carmona-Fontaine C, Matthews HK, Kuriyama S, Moreno M, Dunn GA, Parsons M, Stern CD, Mayor R (2008). Contact inhibition of locomotion in vivo controls neural crest directional migration. *Nature* 456(7224): 957–961.

Chen YH, Ishii M, Sun J, Sucov HM, Maxson RE Jr. (2007). Msx1 and Msx2 regulate survival of secondary heart field precursors and post-migratory proliferation of cardiac neural crest in the outflow tract. *Dev. Biol.* 308(2): 421–437.

Choudhary B, Ito Y, Makita T, Sasaki T, Chai Y, Sucov HM (2006). Cardiovascular malformations with normal smooth muscle differentiation in neural crest-specific type II TGFbeta receptor (Tgfbr2) mutant mice. *Dev. Biol.* 289(2): 420–429.

Clouthier DE, Hosoda K, Richardson JA, Williams SC, Yanagisawa H, Kuwaki T, Kumada M, Hammer RE, Yanagisawa M (1998). Cranial and cardiac neural crest defects in endothelin-A receptor-deficient mice. *Development* 125(5): 813–824.

Clouthier DE, Williams SC, Hammer RE, Richardson JA, Yanagisawa M (2003). Cell-autonomous and nonautonomous actions of endothelin-A receptor signaling in craniofacial and cardiovascular development. *Dev. Biol.* 261(2): 506–519.

Conway SJ, Henderson DJ, Copp AJ (1997). Pax3 is required for cardiac neural crest migration in the mouse: Evidence from the splotch (Sp2H) mutant. *Development* 124(2): 505–514.

Conway SJ, Henderson DJ, Kirby ML, Anderson RH, Copp AJ (1997). Development of a lethal congenital heart defect in the splotch (Pax3) mutant mouse. *Cardiovasc Res.* 36(2): 163–173.

Conway SJ, Bundy J, Chen J, Dickman E, Rogers R, Will BM (2000). Decreased neural crest stem cell expansion is responsible for the conotruncal heart defects within the splotch (Sp(2H))/Pax3 mouse mutant. *Cardiovasc. Res.* 47(2): 314–328.

Danielian PS, Muccino D, Rowitch DH, Michael SK, McMahon AP (1998). Modification of gene activity in mouse embryos in utero by a tamoxifen-inducible form of Cre recombinase. *Curr. Biol.* 8(24): 1323–1326.

Epstein DJ, Vekemans M, Gros P (1991). Splotch (Sp2H), a mutation affecting development of the mouse neural tube, shows a deletion within the paired homeodomain of Pax-3. *Cell* 67(4): 767–774.

Epstein JA, Li J, Lang D, Chen F, Brown CB, Jin F, Lu MM, Thomas M, Liu E, Wessels A, Lo CW (2000). Migration of cardiac neural crest cells in Splotch embryos. *Development* 127(9): 1869–1878.

Farrell MJ, Burch JL, Wallis K, Rowley L, Kumiski D, Stadt H, Godt RE, Creazzo TL, Kirby ML (2001). FGF-8 in the ventral pharynx alters development of myocardial calcium transients after neural crest ablation. *J. Clin. Invest.* 107(12): 1509–1517.

Feiner L, Webber AL, Brown CB, Lu MM, Jia L, Feinstein P, Mombaerts P, Epstein JA, Raper JA (2001). Targeted disruption of semaphorin 3C leads to persistent truncus arteriosus and aortic arch interruption. *Development* 128(16): 3061–3070.

Franz T (1989). Persistent truncus arteriosus in the Splotch mutant mouse. *Anat. Embryol. (Berl).* 180(5): 457–464.

Gans C, Northcutt RG (1983). Neural crest and the origin of vertebrates: A new head. *Science* 220(4594): 268–273.

Gao Z, Kim GH, Mackinnon AC, Flagg AE, Bassett B, Earley JU, Svensson EC (2010). Ets1 is required for proper migration and differentiation of the cardiac neural crest. *Development* 137(9): 1543–1551.

Goddeeris MM, Schwartz R, Klingensmith J, Meyers EN (2007). Independent requirements for Hedgehog signalling by both the anterior heart field and neural crest cells for outflow tract development. *Development* 134(8): 1593–1604.

Henderson DJ, Copp AJ (1997). Role of the extracellular matrix in neural crest cell migration. *J. Anat.* 191(Pt 4): 507–515.

Hildreth V, Webb S, Bradshaw L, Brown NA, Anderson RH, Henderson DJ (2008). Cells migrating from the neural crest contribute to the innervation of the venous pole of the heart. *J Anat.* 212(1): 1–11.

Hildreth V, Anderson RH, Henderson DJ (2009). Autonomic innervation of the developing heart: Origins and function. *Clin. Anat.* 22(1): 36–46.

Huang GY, Wessels A, Smith BR, Linask KK, Ewart JL, Lo CW (1998) Alteration in connexin 43 gap junction gene dosage impairs conotruncal heart development. *Dev. Biol.* 198(1): 32–44.

Hutson MR, Zhang P, Stadt HA, Sato AK, Li YX, Burch J, Creazzo TL, Kirby ML (2006). Cardiac arterial pole alignment is sensitive to FGF8 signaling in the pharynx. *Dev. Biol.* 295(2): 486–497.

Hutson MR, Kirby ML (2007). Model systems for the study of heart development and disease. Cardiac neural crest and conotruncal malformations. *Semin. Cell Dev. Biol.* 18(1): 101–110.

Hutson MR, Sackey FN, Lunney K, Kirby ML (2009). Blocking hedgehog signaling after ablation of the dorsal neural tube allows regeneration of the cardiac neural crest and rescue of outflow tract septation. *Dev. Biol.* 335(2): 367–373.

Ito K, Sieber-Blum M (1991). In vitro clonal analysis of quail cardiac neural crest development. *Dev. Biol.* 148(1): 95–106.

Jain R, Engleka KA, Rentschler SL, Manderfield LJ, Li L, Yuan L, Epstein JA (2011). Cardiac neural crest orchestrates remodeling and functional maturation of mouse semilunar valves. *J. Clin. Invest.* 121(1): 422–30.

Jerome LA, Papaioannou VE (2001). DiGeorge syndrome phenotype in mice mutant for the T-box gene, Tbx1. *Nat. Genet.* 27(3): 286–291.

Jiang X, Rowitch DH, Soriano P, McMahon AP, Sucov HM (2000). Fate of the mammalian cardiac neural crest. *Development* 127(8): 1607–1616.

Kaartinen V, Dudas M, Nagy A, Sridurongrit S, Lu MM, Epstein JA (2004). Cardiac outflow tract defects in mice lacking ALK2 in neural crest cells. *Development* 131(14): 3481–3490.

Kelly RG, Brown NA, Buckingham ME (2001). The arterial pole of the mouse heart forms from Fgf10-expressing cells in pharyngeal mesoderm. *Dev. Cell.* 1(3): 435–440.

Kempf H, Linares C, Corvol P, Gasc JM (1998). Pharmacological inactivation of the endothelin type A receptor in the early chick embryo: A model of mispatterning of the branchial arch derivatives. *Development* 125(24): 4931–4941.

Kirby ML, Gale TF, Stewart DE (1983). Neural crest cells contribute to normal aorticopulmonary septation. *Science* 220(4601): 1059–1061.

Kretz M, Eckardt D, Krüger O, Kim JS, Maurer J, Theis M, van Rijen HV, Schorle H, Willecke K (2006). Normal embryonic development and cardiac morphogenesis in mice with Wnt1-Cre-mediated deletion of connexin43. *Genesis* 44(6): 269–276.

Kulesa PM, Teddy JM, Stark DA, Smith SE, McLennan R (2008). Neural crest invasion is a spatially-ordered progression into the head with higher cell proliferation at the migratory front as revealed by the photoactivatable protein, KikGR. *Dev. Biol.* 316(2): 275–287.

Kurihara Y, Kurihara H, Oda H, Maemura K, Nagai R, Ishikawa T, Yazaki Y (1995). Aortic arch malformations and ventricular septal defect in mice deficient in endothelin-1. *J Clin. Invest.* 96(1): 293–300.

Lescroart F, Kelly RG, Le Garrec JF, Nicolas JF, Meilhac SM, Buckingham M (2010). Clonal analysis reveals common lineage relationships between head muscles and second heart field derivatives in the mouse embryo. *Development* 137(19): 3269–3279.

Lindsay EA, Vitelli F, Su H, Morishima M, Huynh T, Pramparo T, Jurecic V, Ogunrinu G, Sutherland HF, Scambler PJ, Bradley A, Baldini A (2001). Tbx1 haploinsufficieny in the DiGeorge syndrome region causes aortic arch defects in mice. *Nature* 410(6824): 97–101.

Liu S, Liu F, Schneider AE, St Amand T, Epstein JA, Gutstein DE (2006). Distinct cardiac malformations caused by absence of connexin 43 in the neural crest and in the non-crest neural tube. *Development* 133(10): 2063–2073.

Luo Y, High FA, Epstein JA, Radice GL (2006). N-cadherin is required for neural crest remodeling of the cardiac outflow tract. *Dev. Biol.* 299(2): 517–528.

Mellott DO, Burke RD (2008). Divergent roles for Eph and ephrin in avian cranial neural crest. *BMC Dev. Biol.* 8: 56.

Merscher S, Funke B, Epstein JA, Heyer J, Puech A, Lu MM, Xavier RJ, Demay MB, Russell RG, Factor S, Tokooya K, Jore BS, Lopez M, Pandita RK, Lia M, Carrion D, Xu H, Schorle H, Kobler JB, Scambler P, Wynshaw-Boris A, Skoultchi AI, Morrow BE, Kucherlapati R (2001). TBX1 is responsible for cardiovascular defects in velo-cardio-facial/DiGeorge syndrome. *Cell* 104(4): 619–629.

Mjaatvedt CH, Nakaoka T, Moreno-Rodriguez R, Norris RA, Kern MJ, Eisenberg CA, Turner D, Markwald RR (2001). The outflow tract of the heart is recruited from a novel heart-forming field. *Dev Biol.* 238(1): 97–109.

Moorman AF, Christoffels VM, Anderson RH, van den Hoff MJ (2007). The heart-forming fields: One or multiple? Philos. Trans. R Soc. Lond. *B Biol. Sci.* 362(1484): 1257–1265.

Morrison-Graham K, Schatteman GC, Bork T, Bowen-Pope DF, Weston JA (1992). A PDGF receptor mutation in the mouse (Patch) perturbs the development of a non-neuronal subset of neural crest-derived cells. *Development* 115(1): 133–142.

Nakamura H (1982). Mesenchymal derivatives from the neural crest. *Arch. Histol. Jpn.* 45(2): 127–138.

Nakamura T, Colbert MC, Robbins J (2006). Neural crest cells retain multipotential characteristics in the developing valves and label the cardiac conduction system. *Circ. Res.* 98(12): 1547–1554.

Nathan E, Monovich A, Tirosh-Finkel L, Harrelson Z, Rousso T, Rinon A, Harel I, Evans SM, Tzahor E (2008). The contribution of Islet1-expressing splanchnic mesoderm cells to distinct branchiomeric muscles reveals significant heterogeneity in head muscle development. *Development* 135(4): 647–657.

Niederreither K, Vermot J, Schuhbaur B, Chambon P, Dollé P (2000). Retinoic acid synthesis and hindbrain patterning in the mouse embryo. *Development* 127(1): 75–85.

Pérez-Pomares JM, González-Rosa JM, Muñoz-Chápuli R (2009). Building the vertebrate heart — an evolutionary approach to cardiac development. *Int. J. Dev. Biol.* 53(8–10): 1427–1443.

Perris R, Perissinotto D (2000). Role of the extracellular matrix during neural crest cell migration. *Mech. Dev.* 95(1–2): 3–21.

Phillips HM, Murdoch JN, Chaudhry B, Copp AJ, Henderson DJ (2005). Vangl2 acts via RhoA signaling to regulate polarized cell movements during development of the proximal outflow tract. *Circ. Res.* 96(3): 292–299.

Plotkin LI, Manolagas SC, Bellido T (2002). Transduction of cell survival signals by connexin-43 hemichannels. *J. Biol. Chem.* 277(10): 8648–8657.

Poelmann RE, Mikawa T, Gittenberger-de Groot AC (1998). Neural crest cells in outflow tract septation of the embryonic chicken heart: Differentiation and apoptosis. *Dev. Dyn.* 212(3): 373–384.

Randall V, McCue K, Roberts C, Kyriakopoulou V, Beddow S, Barrett AN, Vitelli F, Prescott K, Shaw-Smith C, Devriendt K, Bosman E, Steffes G, Steel KP, Simrick S, Basson MA, Illingworth E, Scambler PJ (2009). Great vessel development requires biallelic expression of Chd7 and Tbx1 in pharyngeal ectoderm in mice. *J. Clin. Invest.* 119(11): 3301–3310. doi:10.1172/JCI37561.

Reaume AG, de Sousa PA, Kulkarni S, Langille BL, Zhu D, Davies TC, Juneja SC, Kidder GM, Rossant J (1995). Cardiac malformation in neonatal mice lacking connexin43. *Science* 267(5205): 1831–1834.

Richarte AM, Mead HB, Tallquist MD (2007). Cooperation between the PDGF receptors in cardiac neural crest cell migration. *Dev. Biol.* 306(2): 785–796.

Robinson V, Smith A, Flenniken AM, Wilkinson DG (1997). Roles of Eph receptors and ephrins in neural crest pathfinding. *Cell Tissue Res.* 290(2): 265–274.

Sanlaville D, Etchevers HC, Gonzales M, Martinovic J, Clément-Ziza M, Delezoide AL, Aubry MC, Pelet A, Chemouny S, Cruaud C, Audollent S, Esculpavit C, Goudefroye G, Ozilou C, Fredouille C, Joye N, Morichon-Delvallez N, Dumez Y, Weissenbach J, Munnich A, Amiel J, Encha-Razavi F, Lyonnet S, Vekemans M, Attié-Bitach T (2006). Phenotypic spectrum of CHARGE syndrome in fetuses with CHD7 truncating mutations correlates with expression during human development. *J. Med. Genet.* 43(3): 211–217.

Sauka-Spengler T, Bronner-Fraser M (2008). A gene regulatory network orchestrates neural crest formation. *Nat. Rev. Mol. Cell Biol.* 9(7): 557–568.

Schnetz MP, Handoko L, Akhtar-Zaidi B, Bartels CF, Pereira CF, Fisher AG, Adams DJ, Flicek P, Crawford GE, Laframboise T, Tesar P, Wei CL, Scacheri PC (2010). CHD7 targets active gene enhancer elements to modulate ES cell-specific gene expression. *PLoS Genet.* 6(7): e1001023.

Scholl AM, Kirby ML (2009). Signals controlling neural crest contributions to the heart. Wiley Interdiscip. *Rev. Syst. Biol. Med.* 1(2): 220–227.

Sieber-Blum M, Ito K (1995). In vitro clonal analysis of cardiac outflow tract mesenchyme. *Ann. N Y Acad. Sci.* 752: 92–100.

Siebert JR, Graham JM Jr, MacDonald C (1985). Pathologic features of the CHARGE association: Support for involvement of the neural crest. *Teratology* 31(3): 331–336.

Smith A, Robinson V, Patel K, Wilkinson DG (1997). The EphA4 and EphB1 receptor tyrosine kinases and ephrin-B2 ligand regulate targeted migration of branchial neural crest cells. *Curr. Biol.* 7(8): 561–570.

Stains JP, Civitelli R (2005). Gap junctions regulate extracellular signal-regulated kinase signaling to affect gene transcription. *Mol. Biol. Cell.* 16(1): 64–72.

Stolfi A, Gainous TB, Young JJ, Mori A, Levine M, Christiaen L (2010). Early chordate origins of the vertebrate second heart field. *Science* 329(5991): 565–568.

Stottmann RW, Choi M, Mishina Y, Meyers EN, Klingensmith J (2004). BMP receptor IA is required in mammalian neural crest cells for development of the cardiac outflow tract and ventricular myocardium. *Development* 131(9): 2205–2218.

Sullivan R, Huang GY, Meyer RA, Wessels A, Linask KK, Lo CW (1998). Heart malformations in transgenic mice exhibiting dominant negative inhibition of gap junctional communication in neural crest cells. *Dev. Biol.* 204(1): 224–234.

Suzuki HR, Kirby ML (1997). Absence of neural crest cell regeneration from the postotic neural tube. *Dev. Biol.* 184(2): 222–233.

Tallquist MD, Soriano P (2003). Cell autonomous requirement for PDGFRalpha in populations of cranial and cardiac neural crest cells. *Development* 130(3): 507–518.

Tamura Y, Matsumura K, Sano M, Tabata H, Kimura K, Ieda M, Arai T, Ohno Y, Kanazawa H, Yuasa S, Kaneda R, Makino S, Nakajima K, Okano H,

Fukuda K (2011). Neural crest-derived stem cells migrate and differentiate into cardiomyocytes after myocardial infarction. *Arterioscler Thromb. Vasc. Biol.* 31(3): 582–589.

Teddy JM, Kulesa PM (2004). In vivo evidence for short- and long-range cell communication in cranial neural crest cells. *Development* 131(24): 6141–6151.

ten Dijke P, Arthur HM (2007). Extracellular control of TGFbeta signalling in vascular development and disease. *Nat. Rev. Mol. Cell Biol.* 8(11): 857–869.

Tomita Y, Matsumura K, Wakamatsu Y, Matsuzaki Y, Shibuya I, Kawaguchi H, Ieda M, Kanakubo S, Shimazaki T, Ogawa S, Osumi N, Okano H, Fukuda K (2005). Cardiac neural crest cells contribute to the dormant multipotent stem cell in the mammalian heart. *J. Cell Biol.* 170(7): 1135–1146.

Toyofuku T, Yoshida J, Sugimoto T, Yamamoto M, Makino N, Takamatsu H, Takegahara N, Suto F, Hori M, Fujisawa H, Kumanogoh A, Kikutani H (2008). Repulsive and attractive semaphorins cooperate to direct the navigation of cardiac neural crest cells. *Dev. Biol.* 321(1): 251–262.

Trainor PA, Ariza-McNaughton L, Krumlauf R (2002). Role of the isthmus and FGFs in resolving the paradox of neural crest plasticity and prepatterning. *Science* 295(5558): 1288–1291.

Vallejo-Illarramendi A, Zang K, Reichardt LF (2009). Focal adhesion kinase is required for neural crest cell morphogenesis during mouse cardiovascular development. *J. Clin. Invest.* 119(8): 2218–2230.

van den Hoff MJ, Moorman AF, Ruijter JM, Lamers WH, Bennington RW, Markwald RR, Wessels A (1999). Myocardialization of the cardiac outflow tract. *Dev. Biol.* 212(2): 477–490.

Vitelli F, Morishima M, Taddei I, Lindsay EA, Baldini A (2002). Tbx1 mutation causes multiple cardiovascular defects and disrupts neural crest and cranial nerve migratory pathways. *Hum. Mol. Genet.* 11(8): 915–922.

Waldo K, Zdanowicz M, Burch J, Kumiski DH, Stadt HA, Godt RE, Creazzo TL, Kirby ML (1999). A novel role for cardiac neural crest in heart development. *J. Clin. Invest.* 103(11): 1499–1507.

Waldo KL, Kumiski DH, Wallis KT, Stadt HA, Hutson MR, Platt DH, Kirby ML (2001). Conotruncal myocardium arises from a secondary heart field. *Development* 128(16): 3179–3188.

Wang J, Nagy A, Larsson J, Dudas M, Sucov HM, Kaartinen V (2006). Defective ALK5 signaling in the neural crest leads to increased postmigratory neural crest cell apoptosis and severe outflow tract defects. *BMC Dev. Biol.* 6: 51.

Xu X, Francis R, Wei CJ, Linask KL, Lo CW (2006). Connexin 43-mediated modulation of polarized cell movement and the directional migration of cardiac neural crest cells. *Development* 133(18): 3629–3639.

Yanagisawa H, Hammer RE, Richardson JA, Williams SC, Clouthier DE, Yanagisawa M (1998). Role of Endothelin-1/Endothelin-A receptor-mediated signalling pathway in the aortic arch patterning in mice. *J Clin. Invest.* 102(1): 22–33.

Zhang J, Chang JY, Huang Y, Lin X, Luo Y, Schwartz RJ, Martin JF, Wang F (2010). The FGF-BMP signalling axis regulates outflow tract valve primordium formation by promoting cushion neural crest cell differentiation. *Circ. Res.* 107(10): 1209–1219.

2

DUAL BUT CONVERGING ROLES: A TALE OF TWO CRESTS

Michael Olaopa and Simon J Conway

Developmental Biology and Neonatal Medicine Program, Herman B Wells Center for Pediatric Research, Indiana University School of Medicine, Indianapolis, IN 46202, USA

The aim of this chapter is to explore the formation, specification, and contribution of pre- and post-migratory neural crest cells to organ development during mammalian embryogenesis. Specifically, we have focused on two key subpopulations — the cardiac neural crest and the trunk neural crest. By reviewing the recent stimulating published data about these two key subpopulations, we hope to shed new light on the importance of these two crest cell types and their distinct but similar roles in providing both structural and functional support to derivatives of the cardiovascular and peripheral nervous systems. We will also highlight the varying decisive roles that each neural crest lineage plays during congenital defect pathogenesis, with emphasis on their functional and structural roles in tissue differentiation and stabilisation.

Introduction

Neural crest (NC) cells are a multipotent and transient migratory lineage originating from the neuroepithelium, which borders the surface ectoderm and gives rise to the embryonic neural tube. NC cells give rise to a staggering array of different cell types, tissues, and organs (Teillet *et al.*, 1999). For this reason, they are often referred to as the fourth germ layer of the developing embryo, in addition to the ectoderm, endoderm, and mesoderm. Further, the post-migratory derivatives of NC are radically diverse in both function and structure. From the cardiovascular to the peripheral nervous system, NC cells have been shown to be absolutely essential during the very early stages of development for normal organogenesis to occur. Specifically, they are important during *in utero* craniofacial, melanocyte, and gastrointestinal development, as well as post-natalas NC stem cells (Dupin *et al.*, 2007; Snider *et al.*, 2007; Adameyko *et al.*, 2009; Sieber-Blum & Hu, 2008).

Different subpopulations of NC have been recognised based on their rostrocaudal position of origin along the axis of the developing embryo, their specific properties, and the nature of their derivatives. The developing neural tube extends in a rostral to caudal direction beginning at the mid-diencephalon. As a result, NC cells have been roughly subdivided into cranial, cardiac, trunk, and sacral NCs. In mouse, the cranial and cardiac NC cells originate from the mid-diencephalon to the fifth somite (Chai *et al.*, 2000; Jiang *et al.*, 2000). Furthermore, the trunk and sacral NC cells begin from the fifth somite to the caudal region of the neural tube (Kirby, 2007). In addition to location, the cranial and trunk NC also differ in functional properties. For instance, the cranial NC is ectomesenchymal, which means that it has the ability to give rise to mesenchymal derivatives, while the trunk NC does not. On the other hand, the trunk NC has the ability to readily regenerate, while the cranial NC has variable capacity for regeneration (Couly *et al.*, 1996; Suzuki & Kirby, 1997). Interestingly, the cardiac NC originates from the mid otic sulcus (the presumptive ear) to the third somite, which represents the caudal region of the cranial NC directly rostral to the beginning of the trunk NC (Creazzo *et al.*, 1998). As a result, the cardiac NC shares some properties of both cranial and trunk NC lineages, including their respective abilities to generate ectomesenchyme and regenerate (Hutson & Kirby, 2007). This chapter focuses primarily on the roles of both cardiac and trunk NCs, and their importance to organ development. There are several steps that cardiac and trunk NC cells undergo in order to properly give rise to this vast array of tissues and organs. As such, there are several series of molecular cascades that need to be spatiotemporally activated at different stages of development resulting in proper NC cell formation, specification and migration, and eventual organ morphogenesis.

NC induction occurs within the neural folds at the neural plate border. In response to signalling from the adjacent non-neural surface ectoderm and neural plate, NCs are specified, proliferate and subsequently undergo epithelial-mesenchymal transformation, prior to emigration from the neural tube and migration along predefined pathways to their target tissues. Cardiac NC cells migrate dorsolaterally out of the neural tube and in between the somites. They migrate toward and eventually colonise the nascent aortic arches, where a subpopulation gives rise to smooth muscle cells. These NC-derived smooth muscle cells surround and stabilise the arches, which will undergo specific asymmetric remodelling required for normal outflow tract (OFT) remodelling. A subset of the cardiac NC cells continues migrating and ultimately colonises the OFT endocardial cushions, wherein they form the aorticopulmonary septum, which divides the two great vessels of the heart — the aorta and the pulmonary artery (Snider *et al.*, 2007; Hutson & Kirby, 2007). Conversely, trunk NC cells migrate dorsoventrally along intersomitic vesicles as well as between the neural tube and somites. A subset of these cells will initially colonise the rostral sclerotome of the somites and eventually contribute to the dorsal root ganglia. However, another subset will continue to migrate past the somites to the region of the dorsal aorta, wherein they will assemble into sympathetic ganglia (Gammill & Roffers-Agarwal, 2010; Huber, 2006). The molecular and genetic regulation of these migratory patterns in both crest populations is tightly controlled and will be discussed in detail later.

Neural crest formation and specification

It is generally believed that the neural plate can induce formation of NC cells via interaction with the adjacent epidermis. Classical experiments done in avian embryos demonstrated effectively that early or late stage neural plate tissue could give rise to NC cells, but that NC formation ceased following neural tube closure along a rostrocaudal developmental gradient (Selleck & Bronner-Fraser, 1995). Furthermore, this induction occurred via interaction between the neural plate and the non-neural ectoderm. Although NC cells originate from a subpopulation of cells located between the developing neural and non-neural ectoderm, studies have shown that these cells can give rise to other cell types as well, indicative of a higher developmental potential (Barembaum & Bronner-Fraser, 2005). Thus, this suggests that signalling from the neural plate is vital in initial induction and formation.

Several families of proteins have been shown to be important in promoting initial NC formation, including the bone morphogenetic proteins (BMP), fibroblast growth factors (FGFs) and Wnt family. Bmp4 is expressed within the epidermis at the border of the neural plate, and several studies have demonstrated that Bmp4 is important in early NC induction. First, Bmp4 has been shown to be absolutely required for cultured neural plates to induce NC, even in the absence of ectoderm (Liem *et al.*, 1995). In addition, beads soaked with Noggin (a dedicated extracellular BMP antagonist), when placed in the region of the epidermis that borders the neural plate leads to abnormal NC expansion (Barembaum & Bronner-Fraser, 2005). In *Xenopus*, FGFs may be required to induce mesoderm, from which NC can ultimately be induced (Monsoro-Burq *et al.*, 2003). However, the exact role of FGF signalling in NC induction is yet to be determined. Some studies have suggested that FGF signalling acts directly through modulation of Noggin to facilitate the early NC induction to occur (Mayor *et al.*, 1997). Furthermore, Wnt activity has been shown to play a role as well. In the absence of Wnt signalling in chick embryos, NC marker expression is diminished or completely eliminated (Garcia-Castro *et al.*, 2002). Similarly, analysis of *Wnt-1/Wnt-3a* double knockout mice indicates that local Wnt signalling regulates the expansion of dorsal NC precursors within the neural tube (Ikeya *et al.*, 1997). It is believed that the initial Wnt expression is required for NC induction and expansion, and then subsequent down-regulation of Wnt must occur in order for the NC cells to delaminate from the neural tube. Interestingly, Wnt down-regulation may also be regulated by a BMP-dependent pathway (Burstyn-Cohen & Kalcheim, 2002). Recent data has also begun to uncover some of the negative regulators that may control the timing of NC specification. Significantly, epigenetic chromatin modifications via histone methylation may maintain precursors in a stem cell-like state prior to their acquisition of characteristic NC migratory ability, and gene expression profiles after neurulation (Strobl-Mazzulla *et al.*, 2010).

The cardiac neural crest lineage

The contribution of NC to the embryonic cardiovascular system was first shown via analysis of chimeric quail and chick embryos (Le Lievre & Le Douarin, 1975). Subsequently, the functional requirement of the NC during cardiovascular development was demonstrated via landmark surgical neural fold ablation studies performed in chick embryos (Kirby *et al.*, 1983). These ablation studies confirmed the importance of cardiac NC cells to normal cardiovascular development, based on a wide range of cardiac defects that occurred following their removal at the neural plate stage. These defects encompassed problems with outflow tract looping and septation, including persistent truncus arteriosus (PTA) and double outlet right ventricle (DORV); aortic arch artery remodelling, including interrupted aortic arch; and accompanying concomitant interventricular septal (VSD) defects. Follow up studies using diI-labelling of NC in quail-chick chimeras (Waldo *et al.*, 1996) and mouse lineage markers (Jiang *et al.*, 2000; Waldo *et al.*, 1999) established that a subpopulation of NC contributed to the aortic arch arteries and the aorticopulmonary septum of the heart. As a result, this subpopulation is called the cardiac NC, and was shown to originate between the otic sulcus and the third somite, which corresponds to the rhombomeres 6–8, the transition region arising between the cranial and trunk NC regions. More recent temporal-spatial genetic ablation of NC in the mouse using *cre/loxP* methods has confirmed these chick data, as the NC-ablated mice embryos also display a spectrum of cardiovascular outflow tract defects and aortic arch patterning abnormalities (Porras & Brown, 2008).

The cardiac NC cells, which are of neuroectodermal origin, emigrate from fused neural tube and migrate along a dorsolateral pattern along specific pathways (Creazzo *et al.*, 1998; Hutson & Kirby, 2007; Waldo *et al.*, 1996). In addition to providing all of the parasympathetic innervation of the heart (Kirby *et al.*, 1983), migratory cardiac NC cells also contribute to the smooth muscle cells that surround the caudal aortic arches and pouches, the ectomesenchymal cells that colonise the outflow tract and divide the great vessels (aorta and pulmonary artery), as well as the developing thymus, thyroid, and parathyroid (Griffith *et al.*, 2009). In addition, they may contribute to the semilunar and atrioventricular valves

(Nakamura *et al.*, 2006), and may be important in regulating maturation of the cardiac conduction system (Poelmann & Gittenberger-de Groot, 1999). However, these lineage mapping results are yet to be tested functionally.

Migratory pathway

The cardiac NC gives rise to the caudal aortic pouches and their respective arteries, as well as the distal region of the outflow tract (Kirby *et al.*, 1983; Waldo *et al.*, 1996). During development, they delaminate from the neural tube and follow a well-defined dorsolateral pathway to their final sites of colonisation, which is discussed in detail later. However, summarising, the cardiac NC will emigrate from the neural tube, through somites 1–4, after which a subset will colonise the aortic arch arterial network (Creazzo *et al.*, 1998), before invading the developing heart in the distal region of the outflow tract (Jiang *et al.*, 2000). As a result, molecular factors important in regulating their migratory pathway, when perturbed, lead to defects in the formation of derivatives from these presumptive structures (Snider *et al.*, 2007; Hutson & Kirby, 2007).

Induction

Although the processes of NC induction have been well studied, the manner in which specification of cardiac NC occurs is still less known. However, recent studies suggest that some of the same molecular factors important in early NC induction (such as Wnt and BMP) may also play a role after cardiac NC cells have emigrated from the neural tube. For instance, Wnt3a has been found to be important in regulating a cardiac-specific gene (*Crip2*) during heart development in the zebrafish (Sun *et al.*, 2008). In the absence of Wnt3a, Crip2 expression within the premigratory NC is perturbed, leading to a delay in cardiac NC cell migration. Furthermore, this effect appears to be rhombomere-specific, as only cardiac NC cells originating from rhombomere 6 are affected.

In both the mouse and chicken, BMP signalling is important in early NC specification (Liem *et al.*, 1995). However, BMP may also be important in regulating a subset of cardiac NC cells as they undergo migration.

BMP-dependent neural crest cells have been shown to be important during OFT remodelling. In mouse embryos where BMP was inhibited, cardiac NC migration was impaired leading to various degrees of outflow tract defects including double-outlet right ventricle and persistent truncus arteriosus (Ohnemus *et al.*, 2002; Tang *et al.*, 2010). The homeobox gene *Lbx1* has also been implicated in regulating cardiac NC specification and migration (Schafer *et al.*, 2003). Lbx1 expression is restricted to a subset of cardiac NC cells; as a result, no migratory defects were discerned in Lbx1-deficient mouse embryos. However, these mutant embryos displayed defects in heart looping and myocardial hyperplasia. These defects are similar to those observed in NC-ablated chick embryos (Waldo *et al.*, 1999). More interestingly, Lbx1 is upregulated in *Pax3* mutant embryos, which are known to exhibit defects in OFT septation (Conway *et al.*, 2000; Conway *et al.*, 2003; Epstein *et al.*, 2000).

Colonisation of aortic arch arteries

Aortic arch arteries develop from endothelial strands within the arches, and connect the aortic sac and the dorsal aortas. In chick embryos, they begin as six symmetric pairs of arteries attached to the paired dorsal aorta. However, in mammals, there are only five symmetric pairs, which are remodelled to give rise to a separate ascending aorta. Elegant studies have clearly demonstrated that the cranial NC gives rise to aortic arches 1 and 2 (Kuratani & Kirby, 1992), while the cardiac NC cells migrate and colonise 3rd, 4th, and 6th aortic arches (Miyagawa-Tomita *et al.*, 1991). This occurs on their way to colonising the aorticopulmonary septum of the OFT, where they coalesce into ridges of connective tissue (Jiang *et al.*, 2000; Creazzo *et al.*, 1998; Hutson & Kirby, 2007). Interestingly, the cardiac NC cells are not required for the initial formation of the aortic arches, but are thought to be important for their final patterning (Jiang *et al.*, 2000; Waldo *et al.*, 1996) and stabilisation (Bradshaw *et al.*, 2009; Snider & Conway, 2007). The cardiac NC cells migrate into each arch as it successively forms and then surround the endothelial cells that form the emerging arches. The role of cardiac NC in stabilising the arches is a key function. In fact, stability of target tissues and vessels appears to be a common function of both cardiac and trunk NC, which is discussed in detail later.

In order for proper aortic arch remodelling to occur, the arteries must undergo a series of asymmetric regression and persistence. The 3rd arch arteries give rise to common carotid arteries, 4th arch arteries contribute to the formation of the distal part of the aortic arch, the brachiocephalic artery and a proximal part of the right subclavian artery, while the left 6th arch arteries contribute to the ductus arteriosus and the proximal parts of the pulmonary arteries (Snider *et al.*, 2007; Kirby, 2007; Conway *et al.*, 2003). Precisely, how the cardiac NC regulates patterning and stabilisation remains unclear, especially as they appear to colonise both the left and right aortic arches equivalently, despite the asymmetric remodelling and persistence of only the left 6th arch arteries.

In addition, the parathyroid and thyroid glands are derived from the endoderm of the 3rd arches via interaction with cardiac NC cells (Rubin *et al.*, 2003). Reciprocal interactions between the migratory cardiac NC cells and the surrounding endothelial cells of the arches have been shown to be important in arch artery remodelling as well. *Endothelin-1* (*Et-1*), which is expressed in the surrounding endothelial cells of the arches but not in the migratory cardiac NC cells, has been shown to be involved in a complex signalling pathway that is required for normal arch remodelling to occur (Yanagisawa *et al.*, 1998). The migratory cardiac NC cells are known to express *Et-A* (Et-1 receptor), and it is believed that proper signalling between the arches and the cardiac NC cells as they migrate through them is important (Creazzo *et al.*, 1998; Hutson & Kirby, 2007) in order to have proper remodelling. Furthermore, mouse embryos, which are deficient for either *Et-1* or *Et-A* display defects in arch remodelling and alignment. Specifically, there is regression of arches 4 and 6, and these mice eventually die from other heart-related defects.

Certain growth factors have also recently been implicated in the remodelling process. Specifically, endothelial-specific expression of *Platelet-derived growth factor alpha* (PDGFα) and activated (phosphorylated) *Vascular endothelial growth factor receptor 2* (VEGFr2) have been shown to be important in the cells surrounding the left 6th arch artery, which continues to persist even as the right 6th arch regresses (Yashiro *et al.*, 2007). This expression of growth factors was shown to be maintained by unilateral distribution of blood flow. It is still unclear what role cardiac NC cells may play in this signalling pathway. However, cardiac NC cell-specific expression of the *receptor* (PDGFRα) has been shown to be important in regulating smooth muscle cell differentiation (Tallquist & Soriano, 2003), which surround the arches and play an important stabilisation role. In addition, asymmetric expression of *Pitx2* in the aortic arch mesoderm has been shown to be important for normal remodelling to occur. Loss of *Pitx2c* results in abnormal remodelling of the aortic arch vessels (Liu *et al.*, 2001), due to poor communication between cardiac NC cells and the endothelial cells surrounding the arteries. This provides another example of the importance of signalling between the cardiac NC cells and the adjacent surrounding endothelial cells for proper arch artery development. Thus, improper specification or migration of cardiac NC cells can lead to abnormal regression or persistence of aortic arches (Hutson & Kirby, 2007), which in some cases results in OFT-related defects.

OFT formation

The OFT is a common vessel exiting the heart, which during development becomes septated into the pulmonary artery and aorta, thereby leading to proper separation of pulmonic and systemic blood flow. Before septation, as a common vessel, it branches at the aortic sac into 3rd, 4th, and 6th aortic arches (Kirby, 2007). Cardiac NC cells are known to play an important role in giving rise to the aorticopulmonary septum, which eventually divides the OFT. Extensive studies on the migration and colonisation of the OFT by cardiac NC cells is now possible with the advent of transgenic mouse *Cre* lines, which allow *Cre/loxP*-mediated lineage traces to be performed. Several transgenic mouse *Cre* lines are now used to study cardiac NC cell migration, including the *P0-Cre, Wnt1-Cre, Pax3-Cre*, and *PlexinA2-Cre* lines (Snider *et al.*, 2007; Jiang *et al.*, 2000; Brown *et al.*, 2001; Lee *et al.*, 1997). Based on these studies, we now have a more comprehensive illustration of the structural contribution of cardiac NC cells and their derivatives to the OFT.

The OFT cushions are initially acellular, and then become filled with cardiac jelly derived from a mixture of mesenchymal cells that migrate from the pharynx and endocardially-derived cells that have undergone epithelial to mesenchymal transformation. Cardiac NC cells also contribute to this mesenchymal population, by migrating via a pair of horseshoe-shaped prongs or condensed streams into the distal region of the OFT. In mouse, the cardiac NC cells do so via a subendocardial route, and they continue all the way to the distal conus of the OFT (Kirby, 2007). During development, this cardiac NC cell-derived mesenchymal population is gradually replaced by myocardium in a process known as myocardialisation (van den Hoff *et al.*, 1999). Myocardialisation of the proximal OFT is important to form the muscular outlet septum of the heart and the atrioventricular canal septum. This process is believed to be regulated by Wnt family of proteins (van den Hoff & Moorman, 2005) via the Wnt-PCP pathway. These studies demonstrated that growth factors such as Wnt5a and Wnt11 are required to be expressed in the OFT cushion mesenchyme in order for normal myocardialisation to occur (Phillips *et al.*, 2005). In addition to the Wnt family, BMP and FGF family of proteins have been shown to be required in conjunction with other factors to induce normal myocardialisation (Somi *et al.*, 2004). The Wnt pathway has also been implicated in regulating the process of cell death of a subset of these cardiac NC cells, which contribute to the conotruncal cushions. After septation is complete, they are believed to eventually undergo apoptosis (Poelmann *et al.*, 1998). These cells transiently express a Wnt receptor *Frizzled-2* (van Gijn *et al.*, 2001), suggesting a role in remodelling and patterning during septation. However, it is believed that some cardiac NC cells can persist even after birth, despite the extensive apoptosis and rearrangement of cardiomyocytes that occur after septation and during myocardialisation (Jiang *et al.*, 2000; Yamauchi *et al.*, 1999).

In addition to septation of the OFT, cardiac NC cells are also important in its elongation and eventual looping and correct alignment. Besides providing a structural framework for the outflow septum, the cardiac NC cells are also believed to be important in regulating cell–cell signalling between the pharynx, during addition of cells derived from the anterior heart field (AHF) to the OFT. In conjunction with these AHF-derived cell lineages, cardiac NC cells play a role in OFT elongation (Black, 2007). The derivatives of the AHF give rise to the myocardial cuff of the OFT, while the cardiac NC cells give rise to the mesenchymal cell population in the cushions. As cells continue to be added to the caudal end of the OFT, the endocardial cushions spiral, leading to rotation of the outflow septum and eventual realignment of the aorta in a posteriolateral direction and the pulmonary artery in an anteriomedial direction (Kirby *et al.*, 1997).

Despite the significant progress that has been made regarding the structural and functional roles of cardiac NC cells in OFT remodelling, elongation, and septation, very little is known about what molecular factors are expressed within these cells post-migration as they undergo differentiation within the OFT cushions. The identification of a post-migratory marker of cardiac NC cells will be both a significant and transformative discovery in the field of cardiovascular development. This could open the door to designing potential therapeutic targets for congenital heart defects and OFT remodelling anomalies associated with aberrant gene expression. Currently, multiple strategies are being employed to address this question, including the use of laser-capture and cell-sorting technologies to isolate and screen expression profiles of cells located in the cardiac NC-derived distal region of the OFT.

Mouse models of congenital heart defects

Numerous animal models have been used to study gene expression and regulation during cardiac NC migration and colonisation. Most of these have been mouse models due to the ease with which genetic manipulation can be carried out in them. One of the first mouse mutants to be characterised was the *Splotch* model, which carries a naturally occurring mutation in the *Pax3* gene. Pax3 is a transcription factor that is involved in numerous developmental processes. In mice, *Pax3* expression begins by embryonic day 8 (Goulding *et al.*, 1991). It is expressed in the dorsal neural tube, including but not restricted to NC progenitors, and somites (musculature) of the developing embryo. It is involved in the formation of myogenic precursors (Epstein *et al.*, 1996), peripheral nervous system and melanocyte differentiation (Galibert *et al.*, 1999), and most importantly, early cardiac NC cell formation and delamination from the neural tube. Its expression diminishes as the cardiac NC cells colonise the arches and it is completely gone as the cardiac NC colonise the OFT (Epstein *et al.*, 2000).

   Splotch mutant embryos exhibit NC-related PTA (Conway *et al.*, 2000; Epstein *et al.*, 2000) and surprisingly all die by embryonic day 14 due to haemodynamic overload and cardiac failure. The pathogenesis of PTA in *Pax3* mutants is due to greatly reduced numbers of migratory NC (Epstein *et al.*, 2000) and the failure of the left 6th aortic arch to persist during remodelling (Conway *et al.*, 2003). The left 6th arch typically gives rise to the pulmonary trunk. Thus, in its absence there is formation of a single OFT from the left 4th arch (normally gives rise to a separate aorta), leading to PTA. However, it is unclear why these *Pax3* mutants fail to survive past the mid-gestation, as Pax3 is not expressed in the heart itself. Furthermore, separation of pulmonic and systemic blood flows *in utero* is not necessary during development as the lungs are not yet functional. It is thought that embryonic lethality is likely due to poor cardiac function and output secondary to myocardial defects (Conway *et al.*, 1997), but it has also been proposed that these defects could be indirectly related to aberrant signalling from fewer migratory cardiac NC cells in the arches and OFT. Transgenic rescue studies using a partial *Pax3*-promoter showed restricted re-expression of *Pax3* within the neural tube and NC in a *Pax3* null background was sufficient to rescue the cardiac phenotype (Li *et al.*, 1999). Not only was PTA rescued, but these mice also survived to birth.

Currently it is believed that *Pax3* is required to suppress another transcription factor, *Msx2* within the neural tube (Kwang *et al.*, 2002). *Msx2* is a homeobox gene and is involved in the regulation of BMP signalling. Mice deficient for *Msx2* are viable, but have cranial NC associated defects including problems with skull ossification, calvarial bones, and teeth formation (Satokata *et al.*, 2000). Interestingly, *Msx2* is upregulated in *Splotch* mutants, and placing the loss of function *Msx2* mutation in the *Pax3* homozygote background rescues the previously described cardiac defects and *in utero* lethality (Kwang *et al.*, 2002). From these studies, it is now clear that the restricted *Pax3* expression within the NT and perhaps the NC specifically, may be important for both normal NC morphogenesis and in maintaining normal cardiac function in the developing mouse embryo. However, the underlying cause of NC-related embryonic lethality in *Pax3* mutants is still unresolved.

Another mouse model of congenital heart disease is the *Sema3C* knockout, which is also expressed in cardiac NC (Feiner *et al.*, 2001). Unlike the *Splotch* mutant mouse models in which initial NC migration is deficient, the *Sema3C* mutant NC cells migrate but fail to invade the OFT cushions. These mutant mice have PTA and interrupted aortic arch, but have no problems with other NC subpopulations. It is believed that signalling between cardiac NC that expresses Sema3C and its surrounding environment, which expresses its receptor Plexin-A2, is important in directing cardiac NC migration along specific pathways and for colonisation of the OFT cushions (Brown *et al.*, 2001; Toyofuku *et al.*, 2007).

Indeed, there are several examples of complementary expression of proteins and their receptors in cardiac NC cells and their surrounding environment respectively (i.e. Vegf and Neuropilin-1, Sdf1 and CxcR4, Sema3C, and Plexin-A2). In fact, a particularly intriguing case is that of retinoic acid and its role in cardiac NC migration and differentiation. Retinoic acid is the active form of vitamin A, and either an excess or suppression of levels affects cardiac development. Retinoic acid functions via its ligand-activated receptors that work as transcription factors. These receptors include the *Retinoic acid receptors* (RARs) and *Retinoic X receptors* (RXRs), which function as both heterodimers and homodimers to transduce retinoic signalling. Mice deficient for different combinations of these receptors display OFT defects including PTA (Lee *et al.*, 1997). Interestingly, based on *Wnt1-Cre*-mediated lineage tracing, these embryos have no problems with NC migration and colonisation. This result is different than that observed in the *Pax3* and *Sema3C* models, which require normal complementary expression of ligand and receptors in both migratory cardiac NC and their surroundings. This study suggests that cardiac NC may not require normal retinoic receptor function, as they are capable of responding directly to retinoic signalling even in the absence of the receptor (Jiang *et al.*, 2002).

In addition to a targeted deletion of genes, transgenic mouse models have also been used as tools to fate-map or lineage trace migration of cardiac NC cells and their derivatives. The first of such studies used a partial *Connexin43* promoter to drive *lacZ* expression in what appeared to be cardiac NC cells (Lo *et al.*, 1999). The location, patterning, and migration of the *lacZ*-expressing cells were similar to those already identified in quail-chick chimera studies mentioned previously. These included cells that gave rise to the aortic arches and OFT region of the heart. However, some aberrant expression of *lacZ* was also seen in the ventricular myocardium, even though cardiac NC is not known to give rise to this region. This was probably due to the endogenous expression of Cx43 in the myocardium (Stoller & Epstein, 2005). Currently gene reporter systems have been complemented with the use of transgenic *Cre* lines that allow permanent labelling of specific cell types and their derivatives. However, there is still an on-going debate on the accuracy or superiority of one transgenic mouse lineage reporter system over another. This is because caveats such as gene expression, integration site of the transgene in use, and sensitivity of the reporter to *Cre*-mediated recombination, are all potential limitations that investigators in the field must consider in order to have a realistic interpretation of results from these traces.

Human syndromes of congenital heart defects

As cardiac NC cells migrate through the aortic arches and into the heart, they are involved in a host of different cell–cell interaction and signalling cascades. As a result, mis-expression of specific molecular factors increases their susceptibility to undergo aberrant migration or differentiation, which can lead to a number of cardiovascular malformations in humans at birth. Congenital malformations are primarily due to aberrant gene expression, which ultimately leads to poor cardiac output both *in utero* and after birth (Olson, 2006). They are the most common birth defect in humans, present in approximately 1% of all live births and up to 10% of stillbirths (Boneva *et al.*, 2001).

One of the more classic syndromes in humans associated with congenital heart defects is DiGeorge syndrome (DGS). Patients with DGS display the CATCH 22 phenotype (cardiac defects, abnormal facies, thymic hypoplasia, cleft palate, hypocalcaemia; associated with microdeletion on chromosome 22q11). Cardiac defects include interrupted aortic arch, PTA, and other arch artery defects. There are a significantly high proportion of patients with these cardiac defects, who also carry the microdeletion (Goldmuntz *et al.*, 1998). The general prevalence is about 1 in 6,000 births, which makes it a fairly common syndrome. The microdeletion varies from patient to patient, but can encompass as much as a 3-million base-pair deletion, and as many as 30 genes have currently been identified, including *Tbx1*.

Tbx1 is a transcription factor and is important during embryonic development for normal development of the thymus, parathyroid, and aortic arches. It is expressed in the pharyngeal endoderm but not in cardiac NC cells, as such its effect on OFT and aortic arch remodelling is believed to be non-cell autonomous. The direct role that Tbx1 plays in cardiac defects is still to be elucidated. However, it is believed to regulate FGF8 expression in the endoderm (Vitelli *et al.*, 2002), which has been suggested to be important for proper differentiation and function of cardiac NC (Yamagishi *et al.*, 2003).

The trunk NC lineage

The trunk NC is another key NC subpopulation that is of significant importance for *in utero* survival. It is located in the region beginning from the 5th somite and extends to the caudal region of the neural tube. It differs significantly in functional and molecular properties from the cranial NC, but shares some similarities with the more proximally-located cardiac NC including its ability to regenerate. However, unlike the cardiac crest which has an ectomesenchymal potential, it does not share this ability (Kirby, 2007).

Migratory pathway

Trunk NC cells, which give rise to elements of the sympathoadrenal lineage, take the dorsoventral migration route, passing mainly through the anterior region of the somites (Loring & Erickson, 1987). During their migration they are exposed to signals from the somites, the ventral neural tube and the notochord. They aggregate in the vicinity of the dorsal aorta, where they form the primary sympathetic ganglia and undergo neuronal and catecholaminergic differentiation (Morikawa *et al.*, 2007).

Trunk NC cells encounter a variety of transient structures, which undergo their own remodelling, as the cells emigrate from the dorsal neural tube and migrate to the embryo periphery. They do so in three stages — early, mid, and late stage — and these are controlled by a tightly regulated expression of specific molecular signals (Gammill & Roffers-Agarwal, 2010; Huber *et al.*, 2005; Roffers-Agarwal & Gammill, 2009). During early stages of migration, trunk NC cells emigrate ventrally between the neural tube and somites along intersomitic vessels. At this stage, the NC cells express type2 Neuropilin (Nrp2), while the presumptive dermomyotome within the somites express type3A Semaphorin (Sema3A). As this occurs, the epithelial somites will dissociate to give rise to dermomyotome and the eventual sclerotome. This dissociation is believed to be caused by an upregulation of Nrp1 within the trunk NC (Schwarz *et al.*, 2009). This is discussed in further detail later.

During the middle phase of migration, a subset of NC cells migrate dorso- or ventrolaterally (in mouse and chick, respectively) into the rostral sclerotome, while the remaining cells aggregate at the dorsal aorta. Low levels of Sema3A and Sema3F expression in the caudal sclerotome, but mainly Sema3A in the dermomyotome, are thought to be important in repelling the NC cells at this stage from the intersomitic space (Schwarz *et al.*, 2009). This leads to restriction of the NC cells to only the rostral sclerotome, thereby regulating migration routes (Gammill *et al.*, 2006). Finally, the subset of trunk NC cells that invaded the somites aggregate to form the segmented dorsal root ganglia. This is triggered via downregulation of Sema3F in the caudal sclerotome and a corresponding upregulation of Sema3A (Roffers-Agarwal & Gammill, 2009). Alternatively, the subset of trunk NC cells that colonises the dorsal aortae is induced along the sympathoadrenal lineage, and begins to aggregate into the primary sympathetic ganglia. This process of induction is further discussed below.

Induction

There are two major schools of thought regarding trunk NC induction and specification into the sympathoadrenal cell lineage. It has been suggested that both the sympathetic precursors (which give rise to neurons in the sympathetic ganglia) and the chromaffin cells (which give rise to the adrenal lineage), originate from the same homogenous population of cells. However, more recently, it has also been proposed that these two lineages may originate from different precursor pools, which may be important in determining which subpopulations of trunk NC colonise the dorsal aorta and which migrate onwards into the adrenal anlage (Huber, 2006; Reiprich *et al.*, 2008).

A major player in trunk NC induction is the BMP family of proteins (Huber, 2006). It is believed that as NC cells migrate pass the dorsal aorta, they are inducted by BMP signalling to develop into sympathoadrenal precursors. In addition to BMP signalling, expression of other factors such as Gata3, Phox2b, and Mash1/2 have been shown to be important for proper specification of the NC along both the sympathetic and adrenal lineages (Huber *et al.*, 2005). However, in both cases, induction is believed to occur both along the neural tube and migratory pathway, and finally at the site of colonisation at the dorsal aorta. Despite the fact that adrenergic markers such as Tyrosine hydroxylase (Th) are not seen to be expressed until the trunk NC colonise the dorsal aorta, the tissues lining the migratory pathway clearly play a role in inducing trunk NC along the sympathoadrenal lineage (Groves *et al.*, 1995).

It is now believed that the notochord, ventral neural tube and somitic mesoderm are all important in inducing adrenergic differentiation in trunk NC cells. This was demonstrated by a number of classic chick explant experiments (Howard & Bronner-Fraser, 1985) and ablation experiments involving the removal of both notochord and ventral tube, which inhibited differentiation of trunk NC into sympathetic ganglia (Teillet & Le Douarin, 1983). According to similar experiments, which also involved removal of notochord and neural tube, induction at the site of colonisation of the dorsal aorta is still dependent on proper expression of earlier factors along the migratory pathway (Groves *et al.*, 1995).

Based on sophisticated pharmacological rescue experiments within genetic mice mutant models, it has now been revealed that while Hand2 is not required for the initial differentiation of the sympathetic nervous system, it is essential for determining the noradrenergic fate of neurons fated along the sympathoadrenal lineage (Morikawa *et al.*, 2007). These data showed that classical markers of sympathetic neuronal differentiation, such as Phox2, Mash1, and Gata2, were unchanged in mouse embryos where *Hand2* had been ablated within the NC lineage. Thus, it is still unclear how the *Hand2* transcription factor regulates noradrenergic markers such as Th and Dbh, which both exhibited reduced expression levels in *Hand2* mutant embryos.

Colonisation of sympathetic ganglia

The formation and colonisation of sympathetic ganglia occurs in three major stages — migration, de-segmentation, and re-segmentation. Initially, migratory trunk NC cells migrate segmentally from the dorsal somite towards the dorsal aorta. They are guided by a complementary series of receptor expressions and their corresponding ligands. Properly specified trunk NC cells will intrinsically express ErbB2, ErbB3, and CxcR4 receptors (Gammill & Roffers-Agarwal, 2010). These factors provide guidance cues to the NC as they migrate through the dorsal somite, which expresses receptor-specific ligands such as Neuregulin (ligand for ErbB2, ErbB3 receptors) and CxcL12 (ligand for CxcR4 receptor) (Belmadani *et al.*, 2005). In addition to these guidance cues, the dermomyotome and notochord also express Sema3A which is important in repelling trunk NC cells via signalling with Nrp1, which is expressed within the NC themselves (Schwarz *et al.*, 2009). Thus, via combinatorial repulsion and attraction cues, the trunk NC cells are able to migrate through the dorsal somite toward the dorsal aorta. The subsequent stages occur at the site of the dorsal aorta, and involve the segmental status of the migratory trunk NC cells. As they colonise the aorta, they become de-segmented, allowing them to diffuse along the outer wall. At this stage, segmental populations of trunk NC are indistinguishable from each other based solely on their location. However, once the NC are evenly dispersed along the aorta, they undergo resegmentation, enabling formation of the metameric sympathetic ganglia (Schwarz *et al.*, 2009).

Adrenal gland formation

In the mouse, adrenal development begins at E11.5, which is at least two days post-delamination of trunk NC from the neuroepithelium. The adrenal gland is formed from an adrenal primordium located along the urinogenital ridge of the developing embryo. Trunk NC cells are important in forming the medulla of the adrenals, which is surrounded by a multilayered mesoderm-derived cortex. The cortex is formed by budding from the coelomic epithelium between the mesogastrium and the urogenital fold. Alternatively, the medulla is composed of chromaffin cells and ganglion cells, which produce and circulate catecholamines respectively.

During development, trunk NC cells that have been fated to differentiate along the sympathoadrenal lineage migrate from the dorsal neural tube to the adrenal primordium. Certain transcription factors, such as those belonging to group E of the Sox protein family (SoxE), including Sox8,

Sox9, and Sox10, may be important during the process of adrenal gland formation (Reiprich *et al.*, 2008). During migration from the neural tube, as described earlier, trunk NC pass by the anterior region of the somites and a subset of these cells will colonise the sympathetic ganglia located adjacent to the dorsal aorta. In the mouse, by E12.5, the trunk crest has already colonised the adrenal medulla, and formation of both the medulla and cortex is completed by E15 (Ehrhart-Bornstein & Hilbers, 1998).

Expression of Sox8 and Sox9 begins before the trunk NC cells have delaminated from the neural tube, and it is believed to be important in early specification (Cheung *et al.*, 2005; O'Donnell *et al.*, 2006). However as they undergo delamination from the neural tube, Sox9 is downregulated in favor of induction of Sox10, which maintains expression throughout their migration (Lindsley *et al.*, 2007) and eventual colonisation of the adrenal medulla, where it is co-expressed with Sox8 (Deal *et al.*, 2006; Sock *et al.*, 2001). Recent studies suggest that Sox10 is very important for normal adrenal development as Sox10-deficient mice do not form an adrenal medulla, due to lack of trunk NC cells colonising the adrenal anlage(Reiprich *et al.*, 2008). These findings are consistent with earlier studies showing that dominant-negative expression of Sox10 in mice resulted in no adrenal formation (Kapur, 1999). Alternatively, despite expression of Sox8 within the medulla, Sox8-deficient mice did not have significant defects in adrenal formation, except when placed on a Sox10 happloinsufficient background (Reiprich *et al.*, 2008). In addition to regulating trunk NC cell survival, Sox10 may also be important in their differentiation. In a Sox10-deficient environment, trunk NC cells colonise the mouse dorsal aorta in reduced numbers, lack proper specification and eventually undergo apoptosis. According to the studies, these cells lacked proper expression of Phox2b and Mash1, indicative of a previously suggested role of Sox10 in induction of molecular factors important in establishing and maintaining the sympathoadrenal lineage (Kim *et al.*, 2003; Lo *et al.*, 1998).

Mouse models of sympathoadrenal defects

As is the case with congenital heart defects, several mouse models have also been used to study gene expression and regulation during trunk NC migration and colonisation. As a result, we now have a clearer understanding of the importance of receptor-ligand interaction between the trunk NC cells and their surrounding environment during sympathogenesis. A particular ligand of interest is Artemin, which is expressed in peripheral blood vessels and adjacent to sympathetic neuronal projections (Honma *et al.*, 2002). It is important in providing attractive cues for trunk NC cells, and it does so via interaction with its receptor, GFRα3, which is expressed intrinsically in sympathetic precursor NC cells. Mice deficient for both GFRα3 and Artemin have fewer sympathetic neurons, suggesting that these may be important in maintaining proliferation of precursor cells (Honma *et al.*, 2002).

Another important receptor-ligand interaction is that of Nrp1 (expressed in NC cells) and its receptor, Sema3a. Mice deficient for Sema3a either fail to undergo proper re-segmentation into sympathetic ganglia at the dorsal aorta, leading to reduced number of sympathetic neurons (Kawasaki *et al.*, 2002). As mentioned above, Neuregulin and its receptors (ErbB2/3) are important in regulating proper migration and colonisation of the trunk crest at the dorsal aorta. Mice deficient for neuregulin1 have been shown to have hypoplastic primary sympathetic ganglia, while mutant ErbB3 mouse embryos display a deficiency of migratory trunk NC cells and eventually die by mid-gestation. The observed lethality in these embryos was caused by a reduction in catecholamine levels due to a deficient number of neural crest precursor cells (Britsch *et al.*, 1998). While these studies have further elucidated the importance of receptor-ligand signalling in NC proliferation and differentiation, they have also created more unanswered questions such as the exact role of trunk NC specification and differentiation in embryonic survival.

Human syndromes of sympathoadrenal disease

As mentioned previously, congenital malformations are fairly common (1% of live births) and are primarily due to aberrant gene expression. Evident by the diverse range of cell–cell signalling that must occur in order for trunk NC cells to properly colonise the adrenal medulla, it is no surprise that defects in adrenal morphogenesis and function can occur. The adrenal gland is important in regulating a number of endocrine and metabolic processes. As a result, several diseases of the adrenals exist, including congenital adrenal hyperplasia (CAH) and Cushing's syndrome (Bottner & Bornstein, 2001).

CAH is the most common form of adrenal insufficiency in children, occurring in about 1 in 10, 000 live births (Shulman *et al.*, 2007). It results from defects in the production of cortisol from cholesterol, which is mainly caused by a deficiency in 21-hydroxylase enzyme. However, adrenal insufficiency has multiple pathologies, and it is associated with patients that present with other clinical symptoms. For example, mutations in the gene encoding the pituitary transcription factor paired-like homeobox 1 (*PROP1*) has been implicated in adrenal insufficiency (Bornstein, 2009). These patients present with deterioration in their anterior pituitary function, and often require hydrocortisone treatment. To our knowledge, there are not many mouse models that completely phenocopy the clinical symptoms seen in patients. Future studies aimed at delineating the exact role of the trunk NC cell population in regulating adrenal function could be helpful in identifying more effective therapeutic targets.

Functional duality of both crests

Although both the cardiac and trunk NC cells have distinct functions during development and give rise to very different structural derivatives, they also have some overlapping roles in determining embryonic lethality. Based on more recent studies, it is becoming increasingly clear that both NC populations have either a direct or indirect role in regulating output and function of the developing heart, and thus, embryonic survival. This is discussed in detail below.

Embryonic lethality associated with CNC

Myocardial defects associated with cardiac NC have been an interesting conundrum in the field of cardiovascular development. This is primarily because cardiac NC cells are not believed to directly contribute to the myocardium of the developing heart. Quite a few animal models of NC have been found to also have problems with myocardial function. One of the earliest models was studied in chick embryos, in which the NC had been surgically ablated (Kirby *et al.*, 1983). These embryos, in addition to classic NC-related defects, also displayed signs of myocardial dysfunction and died before birth. Specifically, they had depressed calcium transients, problems with excitation–contraction coupling and low ejection fraction (Leatherbury *et al.*, 1990).

In mouse, embryos homozygous for the *Splotch* (*Sp^{2H}*) mutation in the *Pax3* gene display similar conotruncal anomalies associated with cardiac NC malformations (Conway *et al.*, 2000; Conway *et al.*, 2003; Epstein *et al.*, 2000). However, these embryos also die by mid-gestation due to apparent poor cardiac output (Conway *et al.*, 1997), and exhibit similar myocardial-related defects to the chick-ablation model. In *Splotch* embryos, there is a failure of cardiac NC to populate the arches and OFT septum in proper numbers (Bradshaw *et al.*, 2009). However, unlike the ablation model, there are still some cardiac NC-derived cells that make it into the aorticopulmonary septum.

There has been some suggestion that cardiac NC cells are required to mediate growth factor signalling between the OFT myocardium and the ventral pharyngeal endoderm. Specifically, studies have reported FGF signalling to be important for normal elongation and remodelling (Farrell *et al.*, 2001). Continuous expression of FGF2 and FGF8 in the ventral pharyngeal endoderm can be detected as cardiac NC cells are still migrating and before they reach the OFT myocardium. In chick embryos, FGF upregulation occurs prior to the arrival of cardiac NC cells in the OFT. Conversely, happloinsufficiency of *FGF8* in embryonic mice also results in PTA/OFT alignment defects due, in part, to an increase in cardiac NC cell apoptosis (Abu-Issa *et al.*, 2002; Brown *et al.*, 2004). This suggests that regulated FGF8 expression in the pharyngeal endoderm is important in both OFT development and perhaps myocardial stability. Interestingly, reduction of FGF signalling in chick embryos that have undergone cardiac NC ablation is sufficient to normalise abnormal calcium transients reported earlier. Furthermore, back transplantation of cardiac NC cells into NC-ablated chick embryos is able to normalise calcium transients and partially rescue the myocardial phenotype (Farrell *et al.*, 2001). To our knowledge, FGF signalling has not been comprehensively studied in the *Pax3* mutant mouse model. However, it is clear that cardiac NC cells are important in mediating signals to the myocardium of the developing heart, which must be required for normal cardiac function and output.

Embryonic lethality associated with Trunk NC

Similarly, there are a number of animal models of trunk NC that have been shown to be associated with myocardial defects. Perhaps, the most significant model reported is that of the *Tyrosine hydroxylase (Th)* knock-out mice, which are embryonic lethal by E11–15. These mice were grossly normal, but displayed congestion of blood in the liver and atrial wall thinning. Furthermore, they also had slight bradycardia and disorganisation of ventricular cardiomyocytes (Zhou *et al.*, 1995). Since trunk NC do not contribute directly to the cardiomyocyte cell population, the observed cardiac defects proved to be an even more significant result. The authors postulated that the observed lethality was due to a catecholamine deficiency caused by failure of trunk NC to properly give rise to developing sympathetic neurons, adrenal chromaffin cells, and enteric neurons. They confirmed this by feeding pregnant dams with L-DOPA, a catecholamine intermediate, and were able to rescue at least 50% of mutant embryos to term, even though the rescued mutants were severely runted and died within a few weeks after birth. They also observed similar results in *dopamine beta-hydroxylase (Dbh)* knockout mice (Zhou & Palmiter, 1995). Dbh is also involved in later stages of catecholamine synthesis, by catalysing conversion of dopamine to norepinephrine. However, since the observed lethality occurred before cathecolaminergic neurotransmission is completely established, the study postulated that catecholamines probably act in a paracrine manner at these early stages (Zhou *et al.*, 1995; Zhou & Palmiter, 1995; Thomas *et al.*, 1995).

In addition to the *Th* knockout mice, the *Gata3* knockout mouse model has been quite informative in this regard. These mice are also embryonic lethal by mid-gestation, and display deficiencies in Th, Dbh, and noradrenaline (Lim *et al.*, 2000). The mutants were also rescued to birth by administration of DOPS (a synthetic catecholamine intermediate) to the pregnant dams. Similar to the *Th* knockouts, these mutants were also reported to have loosely organised cardiomyocyte structure, as well as blood congestion in the heart, and had poorly developed neural crest-derived structures. However, the authors were unable to conclusively link the exact role of Gata3 in Th/Dbh activity and subsequent heart failure in these mutants. Based on subsequent *in vitro* studies, it is now believed that Gata3 can directly bind to the *Th* promoter and regulate its transcription (Hong *et al.*, 2006).

More recent studies, have suggested that while the expression of Th and Gata3 within the trunk NC may be required for normal sympathetic neuron development, expression of much earlier molecular factors are important in determining their noradrenergic identity. The *Hand2* has been shown to be required for early specification of the noradrenergic cell lineage. Conditional deletion of *Hand2* within the neural crest leads to embryonic lethality by mid-gestation and severe cardiovascular defects (Morikawa *et al.*, 2007). These mutants displayed reduced specific reduction in the expression of Th and Dbh. However, it concluded that *Hand2* regulation of both Th and Dbh expression is independent of Gata2/3, Phox2a/b, and Mash1, which are established markers of sympathetic neuronal differentiation (Pattyn *et al.*, 1999; Sommer *et al.*, 1995; Tsarovina *et al.*, 2004), but are all normally expressed in these mutants.

Thus, despite cardiac and trunk NC cells emanating from different anterior-posterior regions, having distinct functions and giving rise to very different structural derivatives, it would appear that both lineages may functionally interact and be jointly required for *in utero* survival and normal development. Future studies aimed at identifying the extent of cross-talk between NC subpopulations will be needed to fully understand the diverse roles that the multipotent NC plays during embryogenesis.

Acknowledgements

We are grateful to members of the Conway laboratory for their continuing support and insights. These studies were supported, in part, by American Heart Association Pre-doctoral Fellowship (MO), Riley Children's Foundation (SJC), and Indiana University Department of Pediatrics (Neonatal–Perinatal Medicine) and NIH HL60714 grant (SJC).

Table 1. Molecular regulation of cardiac and trunk NC populations during development.

Crest Population	Major Role	Gene	Specific Role	Role-specific Expression Pattern	Species	Reference
Neural Crest	*Premigratory Induction*	Bmp4	NC formation and expansion	Neural plate epidermis	Chicken	(Barembaum and Bronner-Fraser, 2005; Liem et al., 1995)
		FGFs	NC formation and expansion	Mesoderm	Xenopus	(Monsoro-Burq et al., 2003; Mayor et al., 1997)
		Wnts	NC formation and expansion	Dorsal neural tube	Chicken	(Garcia-Castro et al., 2002; Burstyn-Cohen et al., 2002)
Cardiac NC	*Premigratory Induction*	Wnt3a	Cardiac NC cell induction	Rhombomere 6-specific NC cells	Zebrafish	(Sun et al., 2008)
		Crip2	Cardiac NC cell induction	Premigratory cardiac NC cells	Zebrafish	(Sun et al., 2008)
		BMPs	Outflow tract remodelling	Subset of migratory cardiac NC cells	Mouse; Chicken	(Ohnemus et al., 2002; Somi et al., 2004)
		Lbx1	Heart looping, myocardial homeostasis	Subset of migratory cardiac NC cells	Mouse	(Schafer et al., 2003)

(Continued)

Table 1. (*Continued*)

Crest Population	Major Role	Gene	Specific Role	Role-specific Expression Pattern	Species	Reference
	Migration/ Colonisation	*Et-1*	Aortic arch artery remodelling; Attracts ET-A-positive cardiac NC cells towards arches	Endothelial cells surrounding aortic arches	Mouse	(Yanagisawa et al., 1998)
		Et-A	Aortic arch artery remodelling; Guides cardiac NC cells towards ET-1-positive aortic arches	Migratory cardiac NC cells	Mouse	(Yanagisawa et al., 1998)
		Vegfr2	Left 6th arch artery persistence	Endothelial cells surrounding left 6th aortic arch	Mouse	(Yashiro et al., 2007)
		Pdgfα	Left 6th arch artery persistence	Endothelial cells surrounding left 6th aortic arch	Mouse	(Yashiro et al., 2007)
		Pdgfrα	Smooth muscle cell differentiation; Aortic arch artery stabilisation	Migratory cardiac NC cells	Mouse	(Tallquist & Soriano, 2003)

(*Continued*)

Table 1. (*Continued*)

Crest Population	Major Role	Gene	Specific Role	Role-specific Expression Pattern	Species	Reference
		Pitx2c	Aortic arch artery remodelling	Aortic arch mesoderm	Mouse	(Liu *et al.*, 2001)
		Frizzled2	Cardiac NC cell apoptosis; Outflow tract septation	Cardiac NC cells	Mouse; Chicken	(Poelmann *et al.*, 1998; van Gijn *et al.*, 2001)
Trunk NC	*Migration*	*Nrp1*	Guides trunk NC cells away from Sema3A-positive intersomitic space	Migratory trunk NC cells	Mouse	(Schwarz *et al.*, 2009a; Schwarz *et al.*, 2009b; Kawasaki *et al.*, 2002)
		Nrp2	Guides trunk NC cells towards Sema3F-positive sclerotome	Migratory trunk NC cells	Mouse	(Roffers-Agarwal & Gammill, 2009)
		Sema3A	Repels Nrp1-positive trunk NC cells from intersomitic space	Somitic dermomyotome	Mouse	(Schwarz *et al.*, 2009a; Schwarz *et al.*, 2009b; Kawasaki *et al.*, 2002)
		Sema3F	Attracts Nrp2-positive trunk NC cells to sclerotome	Somitic caudal sclerotome	Mouse	(Roffers-Agarwal & Gammill, 2009)

(*Continued*)

Table 1. (*Continued*)

Crest Population	Major Role	Gene	Specific Role	Role-specific Expression Pattern	Species	Reference
	Colonisation	*ErbB2*	Sympathetic ganglia differentiation; Receives guidance cues from Neuregulin-positive dorsal somites	Specified trunk NC cells	Mouse	(Britsch *et al*, 1998)
		ErbB3	Sympathetic differentiation; Receives guidance cues from Neuregulin-positive dorsal somites	Specified trunk NC cells	Mouse	(Britsch *et al*, 1998)
		CxcR4	Sympathetic differentiation; Receives guidance cues from CxcL12-positive dorsal somites	Specified trunk NC cells	Mouse	(Belmadani *et al*, 2005)

(*Continued*)

Table 1. (*Continued*)

Crest Population	Major Role	Gene	Specific Role	Role-specific Expression Pattern	Species	Reference
		Neuregulin	Sympathetic differentiation; Provides guidance cues to ErbB2- and ErbB3-positive trunk NC cells	Dorsal somites	Mouse	(Britsch *et al.*, 1998)
		CxcL12	Sympathetic ganglia differentiation; Provides guidance cues to CxcR4-positive trunk NC cells	Dorsal somites	Mouse	(Belmadani *et al.*, 2005)
	Migratory and Post-migratory Induction	*Bmp4*	Neurogenesis, sympathetic differentiation	Migratory trunk NC cells	Mouse	(Huber, 2006)
		Phox2a	Induction of trunk NC cells fated to become sympathoadrenal precursor cells	Migratory trunk NC cells	Mouse	(Lo *et al.*, 1998)

(*Continued*)

Table 1. (*Continued*)

Crest Population	Major Role	Gene	Specific Role	Role-specific Expression Pattern	Species	Reference
		Phox2b	Induction of trunk NC cells fated to become sympathoadrenal precursor cells	Migratory trunk NC cells	Mouse	(Huber et al., 2005)
		Gata3	Induction of trunk NC cells fated to become sympathoadrenal precursor cells	Migratory trunk NC cells	Mouse	(Huber et al., 2005)
		Mash1	Induction of trunk NC cells fated to become sympathoadrenal precursor cells	Migratory trunk NC cells	Mouse	(Huber et al., 2005; Kim et al., 2003)
		Sox8	Adrenogenesis; formation of adrenal medulla	Premigratory and migratory trunk NC cells	Mouse	(Reiprich et al., 2008; Cheung et al., 2005; O'Donnell et al., 2006)

(*Continued*)

Table 1. (*Continued*)

Crest Population	Major Role	Gene	Specific Role	Role-specific Expression Pattern	Species	Reference
		Sox9	Adrenogenesis; formation of adrenal medulla	Premigratory trunk NC cells	Mouse	(Cheung *et al.*, 2005; O'Donnell *et al.*, 2006)
		Sox10	Adrenogenesis; formation of adrenal medulla	Migratory trunk NC cells	Mouse	(Reiprich *et al.*, 2008; Deal *et al.*, 2006; Sock *et al.*, 2001; Kapur, 1999)
		Hand2	Noradrenergic differentiation	Migratory trunk NC cells	Mouse	(Morikawa *et al.*, 2007)
		Th	Noradrenergic differentiation	Migratory and pre-differentiated trunk NC cells	Mouse	(Morikawa *et al.*, 2007; Groves *et al.*, 1995)
		Dbh	Noradrenergic differentiation	Differentiated trunk NC cells	Mouse	(Morikawa *et al.*, 2007)

References

Abu-Issa R *et al.* (2002). Fgf8 is required for pharyngeal arch and cardiovascular development in the mouse. *Development* 129(19): 4613–4625.

Adameyko I *et al.* (2009). Schwann cell precursors from nerve innervation are a cellular origin of melanocytes in skin. *Cell* 139(2): 366–379.

Barembaum M, Bronner-Fraser M (2005). Early steps in neural crest specification. *Semin Cell Dev. Biol.* 16(6): 642–646.

Belmadani A *et al.* (2005). The chemokine stromal cell-derived factor-1 regulates the migration of sensory neuron progenitors. *J. Neurosci.* 25(16): 3995–4003.

Black BL (2007). Transcriptional pathways in second heart field development. Semin. *Cell Dev. Biol.* 18(1): 67–76.

Boneva RS *et al.* (2001). Mortality associated with congenital heart defects in the United States: Trends and racial disparities, 1979–1997. *Circulation* 103(19): 2376–2381.

Bornstein SR (2009). Predisposing factors for adrenal insufficiency. *N. Engl. J. Med.* 360(22): 2328–2339.

Bottner A, Bornstein SR (2001). Lessons learned from gene targeting and transgenesis for adrenal physiology and disease. *Rev. Endocr. Metab. Disord.* 2(3): 275–287.

Bradshaw L *et al.* (2009). Dual role for neural crest cells during outflow tract septation in the neural crest-deficient mutant Splotch(2H). *J. Anat.* 214(2): 245–257.

Britsch S *et al.* (1998). The ErbB2 and ErbB3 receptors and their ligand, neuregulin-1, are essential for development of the sympathetic nervous system. *Genes. Dev.* 12(12): 1825–1836.

Brown CB *et al.* (2001). PlexinA2 and semaphorin signalling during cardiac neural crest development. *Development* 128(16): 3071–3080.

Brown CB *et al.* (2004). Cre-mediated excision of Fgf8 in the Tbx1 expression domain reveals a critical role for Fgf8 in cardiovascular development in the mouse. *Dev. Biol.* 267(1): 190–202.

Burstyn-Cohen T, Kalcheim C (2002). Association between the cell cycle and neural crest delamination through specific regulation of G1/S transition. *Dev. Cell* 3(3): 383–395.

Chai Y *et al.* (2000). Fate of the mammalian cranial neural crest during tooth and mandibular morphogenesis. *Development* 127(8): 1671–1679.

Cheung M *et al.* (2005). The transcriptional control of trunk neural crest induction, survival, and delamination. *Dev. Cell* 8(2): 179–192.

Conway SJ *et al.* (1997). Neural crest is involved in development of abnormal myocardial function. *J. Mol. Cell Cardiol.* 29(10): 2675–2685.

Conway SJ *et al.* (2000). Decreased neural crest stem cell expansion is responsible for the conotruncal heart defects within the splotch (Sp(2H))/Pax3 mouse mutant. *Cardiovasc Res.* 47(2): 314–328.

Conway SJ *et al.* (2003). What cardiovascular defect does my prenatal mouse mutant have, and why? *Genesis* 35(1): 1–21.

Couly G *et al.* (1996). The regeneration of the cephalic neural crest, a problem revisited: The regenerating cells originate from the contralateral or from the anterior and posterior neural fold. *Development* 122(11): 3393–3407.

Creazzo TL *et al.* (1998). Role of cardiac neural crest cells in cardiovascular development. *Annu Rev. Physiol* 60: 267–286.

Deal KK *et al.* (2006). Distant regulatory elements in a Sox10-beta GEO BAC transgene are required for expression of Sox10 in the enteric nervous system and other neural crest-derived tissues. *Dev. Dyn.* 235(5): 1413–1432.

Dupin E *et al.* (2007). Neural crest progenitors and stem cells. *C R Biol.* 330(6–7): 521–529.

Ehrhart-Bornstein M, Hilbers U (1998). Neuroendocrine properties of adrenocortical cells. *Horm. Metab. Res.* 30(6–7): 436–439.

Epstein JA *et al.* (1996). Pax3 modulates expression of the c-Met receptor during limb muscle development. *Proc. Natl. Acad. Sci. USA* 93(9): 4213–4218.

Epstein JA *et al.* (2000). Migration of cardiac neural crest cells in Splotch embryos. *Development* 127(9): 1869–1878.

Farrell MJ *et al.* (2001). FGF-8 in the ventral pharynx alters development of myocardial calcium transients after neural crest ablation. *J. Clin. Invest.* 107(12): 1509–1517.

Feiner L *et al.* (2001). Targeted disruption of semaphorin 3C leads to persistent truncus arteriosus and aortic arch interruption. *Development* 128(16): 3061–3070.

Galibert MD *et al.* (1999). Pax3 and regulation of the melanocyte-specific tyrosinase-related protein-1 promoter. *J. Biol. Chem.* 274(38): 26894–26900.

Gammill LS *et al.* (2006). Guidance of trunk neural crest migration requires neuropilin 2/semaphorin 3F signalling. *Development* 133(1): 99–106.

Gammill LS, Roffers-Agarwal J (2010). Division of labor during trunk neural crest development. *Dev. Biol.* 344(2): 555–565.

Garcia-Castro MI, Marcelle C, Bronner-Fraser M (2002). Ectodermal Wnt function as a neural crest inducer. *Science* 297(5582): 848–851.

Goldmuntz E *et al.* (1998). Frequency of 22q11 deletions in patients with conotruncal defects. *J. Am. Coll. Cardiol.* 32(2): 492–498.

Goulding MD *et al.* (1991). Pax-3, a novel murine DNA binding protein expressed during early neurogenesis. *EMBO J* 10(5): 1135–1147.

Griffith AV *et al.* (2009). Increased thymus- and decreased parathyroid-fated organ domains in Splotch mutant embryos. *Dev. Biol.* 327(1): 216–227.

Groves AK *et al.* (1995). Differential regulation of transcription factor gene expression and phenotypic markers in developing sympathetic neurons. *Development* 121(3): 887–901.

Hong SJ *et al.* (2006). GATA-3 regulates the transcriptional activity of tyrosine hydroxylase by interacting with CREB. *J. Neurochem.* 98(3): 773–781.

Honma Y *et al.* (2002). Artemin is a vascular-derived neurotropic factor for developing sympathetic neurons. *Neuron* 35(2): 267–282.

Howard MJ, Bronner-Fraser M (1985). The influence of neural tube-derived factors on differentiation of neural crest cells *in vitro*. I. Histochemical study on the appearance of adrenergic cells. *J. Neurosci.* 5(12). 3302–3309.

Huber AB *et al.* (2005). Distinct roles for secreted semaphorin signalling in spinal motor axon guidance. *Neuron* 48(6): 949–964.

Huber K (2006). The sympathoadrenal cell lineage: Specification, diversification, and new perspectives. *Dev. Biol.* 298(2): 335–343.

Huber K *et al.* (2005). The role of Phox2B in chromaffin cell development. *Dev. Biol.* 279(2):501–508.

Hutson MR, Kirby ML (2007). Model systems for the study of heart deveopment and disease. Cardiac neural crest and conotruncal malformations. *Semin Cell Dev. Biol.* 18(1): 101–110.

Ikeya M *et al.* (1997). Wnt signalling required for expansion of neural crest and CNS progenitors. *Nature* 389(6654): 966–970.

Jiang X *et al.* (2000). Fate of the mammalian cardiac neural crest. *Development* 127(8): 1607–1616.

Jiang X *et al.* (2002). Normal fate and altered function of the cardiac neural crest cell lineage in retinoic acid receptor mutant embryos. *Mech. Dev.* 117(1–2): 115–22.

Kapur RP (1999). Early death of neural crest cells is responsible for total enteric aganglionosis in Sox10(Dom)/Sox10(Dom) mouse embryos. *Pediatr. Dev. Pathol.* 2(6): 559–569.

Kawasaki T *et al.* (2002). Requirement of neuropilin 1-mediated Sema3A signals in patterning of the sympathetic nervous system. *Development* 129(3): 671–680.

Kim J *et al.* (2003). SOX10 maintains multipotency and inhibits neuronal differentiation of neural crest stem cells. *Neuron* 38(1): 17–31.

Kirby ML (2007). Cardiac neural crest in evolution and development. *Faseb Journal* 21(5): A89–A89.

Kirby ML *et al.* (1997). Abnormal patterning of the aortic arch arteries does not evoke cardiac malformations. *Dev. Dyn.* 208(1): 34–47.

Kirby ML, Gale TF, Stewart DE (1983). Neural crest cells contribute to normal aorticopulmonary septation. *Science* 220(4601): 1059–1061.

Kuratani SC, Kirby ML (1992). Migration and distribution of circumpharyngeal crest cells in the chick embryo. Formation of the circumpharyngeal ridge and E/C8+ crest cells in the vertebrate head region. *Anat. Rec.* 234(2): 263–280.

Kwang SJ *et al.* (2002). Msx2 is an immediate downstream effector of Pax3 in the development of the murine cardiac neural crest. *Development* 129(2): 527–538.

Le Lievre CS, Le Douarin NM (1975). Mesenchymal derivatives of the neural crest: Analysis of chimaeric quail and chick embryos. *J. Embryol Exp. Morphol.* 34(1): 125–154.

Leatherbury L *et al.* (1990). Microcinephotography of the developing heart in neural crest-ablated chick embryos. *Circulation* 81(3): 1047–1057.

Lee M *et al.* (1997). P0 is constitutively expressed in the rat neural crest and embryonic nerves and is negatively and positively regulated by axons to generate non-myelin-forming and myelin-forming Schwann cells, respectively. *Mol. Cell Neurosci.* 8(5): 336–350.

Li J *et al.* (1999). Transgenic rescue of congenital heart disease and spina bifida in Splotch mice. *Development* 126(11): 2495–2503.

Liem KF Jr *et al.* (1995). Dorsal differentiation of neural plate cells induced by BMP-mediated signals from epidermal ectoderm. *Cell* 82(6): 969–979.

Lim KC *et al.* (2000). Gata3 loss leads to embryonic lethality due to noradrenaline deficiency of the sympathetic nervous system. *Nat. Genet.* 25(2): 209–212.

Lindsley A *et al.* (2007). Identification and characterisation of a novel Schwann and outflow tract endocardial cushion lineage-restricted periostin enhancer. *Dev. Biol.* 307(2): 340–355.

Liu C *et al.* (2001). Regulation of left-right asymmetry by thresholds of Pitx2c activity. *Development* 128(11): 2039–2048.

Lo L *et al.* (1999). Specification of neurotransmitter identity by Phox2 proteins in neural crest stem cells. *Neuron* 22(4): 693–705.

Lo L, Tiveron MC, Anderson DJ (1998). MASH1 activates expression of the paired homeodomain transcription factor Phox2a, and couples pan-neuronal and subtype-specific components of autonomic neuronal identity. *Development* 125(4): 609–620.

Loring JF, Erickson CA (1987). Neural crest cell migratory pathways in the trunk of the chick embryo. *Dev. Biol.* 121(1): 220–236.

Mayor R, Guerrero N, Martinez C (1997). Role of FGF and noggin in neural crest induction. *Dev. Biol.* 189(1): 1–12.

Miyagawa-Tomita S *et al.* (1991). Temporospatial study of the migration and distribution of cardiac neural crest in quail-chick chimeras. *Am J. Anat.* 192(1): 79–88.

Monsoro-Burq AH, Fletcher RB, Harland RM (2003). Neural crest induction by paraxial mesoderm in Xenopus embryos requires FGF signals. *Development* 130(14): 3111–3124.

Morikawa Y *et al.* (2007). Hand2 determines the noradrenergic phenotype in the mouse sympathetic nervous system. *Dev. Biol.* 307(1): 114–126.

Nakamura T, Colbert MC, Robbins J (2006). Neural crest cells retain multipotential characteristics in the developing valves and label the cardiac conduction system. *Circ. Res.* 98(12): 1547–1554.

O'Donnell M *et al.* (2006). Functional analysis of Sox8 during neural crest development in Xenopus. *Development* 133(19): 3817–3826.

Ohnemus S *et al.* (2002). Aortic arch and pharyngeal phenotype in the absence of BMP-dependent neural crest in the mouse. *Mech. Dev.* 119(2): 127–135.

Olson EN (2006). Gene regulatory networks in the evolution and development of the heart. *Science* 313(5795): 1922–1927.

Pattyn A *et al.* (1999). The homeobox gene Phox2b is essential for the development of autonomic neural crest derivatives. *Nature* 399(6734): 366–370.

Phillips HM *et al.* (2005). Vangl2 acts via RhoA signalling to regulate polarised cell movements during development of the proximal outflow tract. *Circ. Res.* 96(3): 292–299.

Poelmann RE, Gittenberger-de Groot AC (1999). A subpopulation of apoptosis-prone cardiac neural crest cells targets to the venous pole: Multiple functions in heart development? *Dev. Biol.* 207(2): 271–286.

Poelmann RE, Mikawa T, Gittenberger-de Groot AC (1998). Neural crest cells in outflow tract septation of the embryonic chicken heart: Differentiation and apoptosis. *Dev. Dyn.*, 212(3): 373–384.

Porras D, Brown CB (2008). Temporal-spatial ablation of neural crest in the mouse results in cardiovascular defects. *Dev. Dyn.* 237(1): 153–162.

Reiprich S *et al.* (2008). SoxE proteins are differentially required in mouse adrenal gland development. *Mol. Biol. Cell* 19(4): 1575–1586.

Roffers-Agarwal J, Gammill LS (2009). Neuropilin receptors guide distinct phases of sensory and motor neuronal segmentation. *Development* 136(11): 1879–1888.

Ruhin B *et al.* 2003. Patterning of the hyoid cartilage depends upon signals arising from the ventral foregut endoderm. *Dev. Dyn.* 228(2): 239–246.

Satokata I *et al.* (2000). Msx2 deficiency in mice causes pleiotropic defects in bone growth and ectodermal organ formation. *Nat. Genet.* 24(4): 391–395.

Schafer K *et al.* (2003).The homeobox gene Lbx1 specifies a subpopulation of cardiac neural crest necessary for normal heart development. *Circ. Res.* 92(1): 73–80.

Schwarz Q *et al.* (2009a). Neuropilin 1 signalling guides neural crest cells to coordinate pathway choice with cell specification. *Proc. Natl. Acad. Sci. USA*, 106(15): 6164–6169.

Schwarz Q *et al.* (2009b). Neuropilin-mediated neural crest cell guidance is essential to organise sensory neurons into segmented dorsal root ganglia. *Development* 136(11): 1785–1789.

Selleck MA, Bronner-Fraser M (1995). Origins of the avian neural crest: The role of neural plate-epidermal interactions. *Development* 121(2): 525–538.

Shulman DI, Palmert MR, Kemp SF (2007). Adrenal insufficiency: still a cause of morbidity and death in childhood. *Pediatrics* 119(2): e484–e494.

Sieber-Blum M, Hu Y (2008). Epidermal neural crest stem cells (EPI-NCSC) and pluripotency. *Stem Cell Rev.* 4(4): 256–260.

Snider P *et al.* (2007). Cardiovascular development and the colonising cardiac neural crest lineage. *Scientific World Journal* 7: 1090–1113.

Snider P, Conway SJ (2007). Developmental biology: The power of blood. *Nature* 450(7167): 180–181.

Sock E *et al.* (2001). Idiopathic weight reduction in mice deficient in the high-mobility-group transcription factor Sox8. *Mol. Cell Biol.* 21(20): 6951–6959.

Somi S *et al.* (2004). Dynamic patterns of expression of BMP isoforms 2, 4, 5, 6, and 7 during chicken heart development. *Anat. Rec. A Discov. Mol. Cell Evol. Biol.* 279(1): 636–651.

Sommer L *et al.* (1995). The cellular function of MASH1 in autonomic neurogenesis. *Neuron* 15(6): 1245–1258.

Stoller JZ, Epstein JA (2005). Cardiac neural crest. Semin. *Cell Dev. Biol.* 16(6): 704–715.

Strobl-Mazzulla PH, Sauka-Spengler T, Bronner-Fraser M (2010). Histone demethylase JmjD2A regulates neural crest specification. *Dev Cell* 19(3): 460–468.

Sun X *et al.* (2008). Wnt3a regulates the development of cardiac neural crest cells by modulating expression of cysteine-rich intestinal protein 2 in rhombomere 6. *Circ. Res.* 102(7): 831–839.

Suzuki HR, Kirby ML (1997). Absence of neural crest cell regeneration from the postotic neural tube. *Dev Biol* 184(2): 222–233.

Tallquist MD, Soriano P (2003). Cell autonomous requirement for PDGFRalpha in populations of cranial and cardiac neural crest cells. *Development* 130(3): 507–518.

Tang S *et al.* 2010. Trigenic neural crest-restricted Smad7 over expression results in congenital craniofacial and cardiovascular defects. *Dev. Biol.* 344(1): 233–247.

Teillet MA, Le Douarin NM (1983). Consequences of neural tube and notochord excision on the development of the peripheral nervous system in the chick embryo. *Dev. Biol.* 98(1): 192–211.

Teillet MA, Ziller C, Le Douarin NM (1999). Quail-chick chimeras. *Methods Mol. Biol.* 97: 305–318.

Thomas SA, Matsumoto AM, Palmiter RD (1995). Noradrenaline is essential for mouse fetal development. *Nature* 374(6523): 643–646.

Toyofuku T *et al.* (2007). Semaphorin-4A, an activator for T-cell-mediated immunity, suppresses angiogenesis via Plexin-D1. *EMBO J* 26(5): 1373–1384.

Tsarovina K *et al.* (2004). Essential role of Gata transcription factors in sympathetic neuron development. *Development* 131(19): 4775–4786.

van den Hoff MJ *et al.* (1999). Myocardialisation of the cardiac outflow tract. *Dev. Biol.* 212(2): 477–490.

van den Hoff MJ, Moorman AF (2005). Wnt, a driver of myocardialisation? *Circ. Res.* 96(3): 274–276.

van Gijn ME *et al.* (2001). Frizzled 2 is transiently expressed in neural crest-containing areas during development of the heart and great arteries in the mouse. *Anat. Embryol. (Berl)* 203(3): 185–192.

Vitelli F *et al.* (2002). Tbx1 mutation causes multiple cardiovascular defects and disrupts neural crest and cranial nerve migratory pathways. *Hum. Mol. Genet.* 11(8): 915–922.

Waldo KL, Kumiski D, Kirby ML (1996). Cardiac neural crest is essential for the persistence rather than the formation of an arch artery. *Dev. Dyn.* 205(3): 281–292.

Waldo KL, Lo CW, Kirby ML (1999). Connexin 43 expression reflects neural crest patterns during cardiovascular development. *Dev. Biol.* 208(2): 307–323.

Yamagishi H *et al.* (2003). Tbx1 is regulated by tissue-specific forkhead proteins through a common Sonic hedgehog-responsive enhancer. *Genes. Dev.* 17(2): 269–281.

Yamauchi Y *et al.* (1999). A novel transgenic technique that allows specific marking of the neural crest cell lineage in mice. *Dev. Biol.* 212(1): 191–203.

Yanagisawa H *et al.* 1998. Role of Endothelin-1/Endothelin-A receptor-mediated signalling pathway in the aortic arch patterning in mice. *J. Clin. Invest.* 102(1): 22–33.

Yashiro K, Shiratori H, Hamada H (2007). Haemodynamics determined by a genetic programme govern asymmetric development of the aortic arch. *Nature* 450(7167): 285–288.

Zhou QY, Palmiter RD (1995). Dopamine-deficient mice are severely hypoactive, adipsic, and aphagic. *Cell* 83(7): 1197–1209.

Zhou QY, Quaife CJ, Palmiter RD (1995). Targeted disruption of the tyrosine hydroxylase gene reveals that catecholamines are required for mouse fetal development. *Nature* 374(6523): 640–643.

3

THE CORNEA, NEURAL CREST AND STEM CELLS

Charles Osei-Bempong[*,†], *Haifa Ali*[*,†]
and Sajjad Ahmad[*,†,‡,§]

*Institute of Genetic Medicine,
Newcastle University, United Kingdom
†North East England Stem Cell Institute,
Newcastle University, United Kingdom
‡Royal Victoria Infirmary, Newcastle upon Tyne,
United Kingdom
§Department of Ophthalmology, Newcastle University,
United Kingdom

The cornea is the clear front of the eye and is composed of three main layers: An epithelium externally, a stroma in the middle, and the endothelium internally. Embryologically, the corneal epithelium develops from the surface ectoderm, while the corneal stroma and endothelium develop from migrating neural crest cells. The corneal epithelium is regenerated by reasonably well recognised stem cells, known as limbal stem cells. There is increasing attention now being given to the stem cells for the corneal stromal fibroblasts or keratocytes. Although a relatively new field in its own right, evidence regarding the presence of neural crest-derived stem cells from the corneal stroma will be discussed by highlighting the major pieces of research in this field.

75

The cornea and its stem cells

The cornea forms a clear avascular structure at the front of the eye (Figure 1). The main functions of the cornea are as a covering for the eye at the front, and to transmit and focus light to the retina for visual perception (Ahmad *et al.*, 2006). Clarity is therefore a vital property of the cornea. The cornea is essentially a three-layered structure: A stratified epithelium on the outside, a thick collagenous stroma containing keratocytes (corneal fibroblasts), and a single layered endothelium (Figure 2). The corneal epithelium is a 5–7 layered stratified epithelium composed of a basal layer of columnar cells, a few layers of wing shaped cells and then more flattened cells on the surface. The corneal epithelium is renewed by stem cells located in a region known as the limbus at the periphery of the cornea (Figure 1). These stem cells for the corneal epithelium are more commonly known as limbal stem cells (Ahmad *et al.*, 2010a; Ahmad

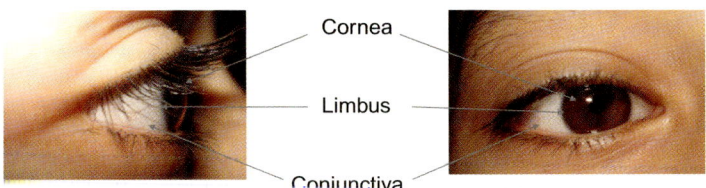

Figure 1.   Front and side views of the human eye highlighting the cornea and limbus.

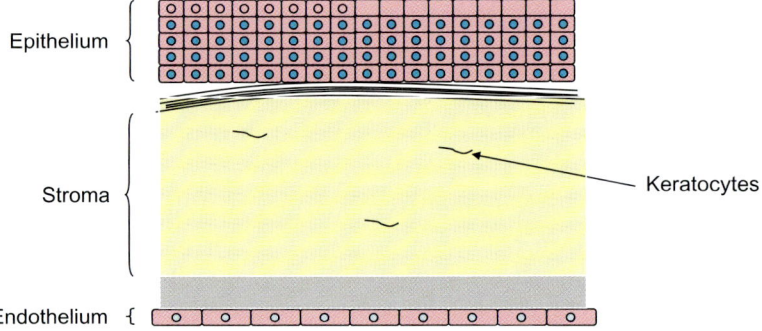

Figure 2.   An illustration indicating the three main layers of the cornea: epithelium, stroma and endothelium.

et al., 2010b). The corneal stroma and endothelium have always been thought of as relatively stable structures with no known stem cells. Despite this conventional thinking, it has recently been described that stem cell-like cells can also be found in the corneal stroma and endothelium (McGowan *et al.*, 2007). The cornea, although avascular, is highly innervated. These corneal nerves are derived from the trigeminal ganglion and terminate as endings in the basal layers of the corneal epithelium.

Embryology of the cornea

Corneal development begins at approximately week 5 in the human embryo (Zieske, 2004). A primitive two-layered epithelium develops from the surface ectoderm covering the developing lens of the eye. Between weeks 5 to 7, this epithelium becomes four cell layers thick and the lens of the developing eye detaches from the ectoderm. Cranial neural crest cells migrate into the space between the epithelium and the lens and form the corneal stromal keratocytes and the endothelium. For corneal stromal development, neural crest cells invade the space between the developing corneal epithelium and endothelium and they become keratoblasts (Hassell & Birk, 2010). Keratoblasts synthesise hyaluronan to form an embryonic stromal extracellular matrix. The keratoblasts then differentiate into stromal keratocytes which synthesise collagen, a major component of the corneal stroma. The trigeminal ganglion and the corneal nerves also develop from the cranial neural crest (Lwigale, 2001).

Evidence for neural crest-derived stem cells in the cornea

Although the corneal stroma was once thought of as a relatively stable structure with low turnover of cells, there is emerging evidence that in fact the corneal stroma and its keratocytes are indeed certainly a source of neural crest-derived stem cells. The evidence for the existence of these stem cells and of their differentiation potential will be discussed here.

Mouse limbus-derived neural crest-derived stem cell-like cells

It has recently been shown by Brandl *et al.* that the postnatal mouse limbus is a source of cells which have fibroblastic morphology, are of neural crest origin, and certainly *in vivo* are multipotent (Brandl *et al.*, 2009). These cells have been termed "neural crest-derived stem cell-like cells" (NCDSCs). These cells were isolated by culturing limbal tissue from mice between postnatal days 1 and 8. Limbal tissue from mice older than this could not give rise to NCDSCs, nor could corneal tissue. These cells are *Sca-1* positive, an important marker for mesenchymal stem cells. They express the neural crest markers *Twist, Slug, Snail,* and *Sox9*. They also express stem cell markers such as *Abcg2* and *Cd34*. To determine the origin of these cells, it was shown that they were positive for corneal stromal keratocyte markers *Lumican* and *Aldh1*, weakly positive for endothelial cell markers such as *Pecam1*, and negative for epithelial cells markers. This data indicates that NCDSCs are most likely to be of stromal origin. Moreover this work shows that NCDSCs *in vitro* are multipotent in nature, differentiating into cells which resemble adipocytes, osteoblasts, and neuronal cells. These studies show that NCDSCs are of neural crest origin, most likely derived from the limbal stroma, and behave like multipotent stem cells. Interestingly these cells have limited *in vitro* proliferative potential, losing it during the first 10 passages of culture, in a similar way to mesenchymal stem cells.

Quail-chick chimeric grafts show multipotency of corneal keratocytes

Lwigale *et al.* used quail/chick chimeric grafts to show two important points: First that the neural crest has an important role in corneal development and secondly that corneal stromal keratocytes possess plasticity and multipotency (Lwigale *et al.*, 2005). In the first set of studies, the dorsal neural tubes of quail embryos were transplanted to chick embryos following ablation of their dorsal neural tubes. The chick embryos were then followed by tracking quail nuclei using immuno-histochemistry until corneal stromal development had occurred in the embryos. These studies showed that the quail-derived neural crest cells migrate to the periocular mesenchymal tissue and then once the developing corneal epithelium and lens have separated, these cells invade the space between the two. The quail-derived neural crest cells can then be seen to develop into the corneal endothelium and also subsequently the corneal stromal cells. Moreover the quail neural crest-derived corneal stromal cells begin to express keratan sulphate which is a proteoglycan found in both developing and adult corneal stroma and shows that the quail-derived cells are functionally active. In these studies, the quail-derived neural crest cells are seen to contribute to the development of the ciliary body of the eye as well as other structures. The second set of studies showed that corneal keratocytes are not terminally differentiated and can display a certain degree of plasticity and multipotency. Corneal stromal keratocytes were isolated from quail embryos and injected into the cranial mesenchyme adjacent to the neural tubes in chick embryos. These studies showed that the injected stromal keratocytes followed the normal neural crest migratory pathways in the chick embryos. Moreover the studies showed that the stromal keratocytes lost their expression of keratan sulphate during migration and this loss of expression was not related to cell division. Subsequently, they also began to express neural crest markers during the migration period. By following the injected quail-derived stromal keratocytes to the eye, the studies showed that these cells contribute to the corneal stroma and endothelium, the iris, ciliary muscles and extraocular muscles. Interestingly, the injected keratocytes also contributed to the development of blood vessels, cardiac mesenchyme and jaw musculature but not neurones, showing that they exhibit a degree of multipotency. This second set of studies is very important as it shows that embryonic stromal keratocytes exhibit a certain degree of plasticity by losing keratan sulphate expression and expressing neural crest markers and morphology. They also have a degree of multipotency being able to form the corneal endothelium, the smooth muscles of blood vessels, the myofibroblasts of the iris and ciliary body and jaw muscles and contribute to the development of the primitive heart.

Role of transforming growth factor β2 in fate specification of neural crest-derived stem cells

Studies by Ittner *et al.* have shown the important role of transforming growth factor β2 (TGFβ2) in inducing periocular neural crest-derived stem cells, originating from the dorsal neural tube in developing mice, to express forkhead/winged-helix factor *Foxc1* and paired like homeodomain factor *Pitx2* (Ittner *et al.*, 2005). These result in corneal endothelial and corneal keratocyte specification respectively. They performed a series of valuable and detailed experiments to show four different aspects of the neural crest's role in eye development. Firstly, they used transgenic mouse studies to show that neural crest-derived cells are found in the anterior periocular mesenchyme giving rise to the corneal endothelium, stromal keratocytes and the eye's chamber angle structures. Interestingly they are also found in developing posterior structures of the eye such as the primary vitreous and cells of the choroid. This shows that neural crest contribution although primarily to anterior eye structures is also important in the posterior aspect of the eye. During eye development there is a peak in *TGFβ2* expression from the lens of the eye. Moreover TGFβ receptor type 2 (*TGFβR2*) is expressed by periocular mesenchymal tissue, lens, primary vitreous, and the retina. The second set of studies by Ittner *et al.* involved neural crest-derived cell specific inactivation of *TGFβr2*. These showed that the mutant mice had small eyes and lacked a normal corneal endothelium, stroma and chamber angles (similar to the features found in Axenfeld–Rieger's anomaly), as well as primary vitreous and retinal anomalies. The third set of studies was again *in vivo* and showed that the loss of TGFβ2 responsiveness in the mutant mice affected the expression of *Foxc1* and *Pitx2*. *Foxc1* localises to the corneal endothelium during normal eye development and failed to be expressed in the mutant mouse corneal endothelium, and indeed these cells underwent apoptosis. *Pitx2* localises to the normally developing corneal stroma and interestingly in these studies although *Pitx2* expression was not seen as expected, the expression of melanocyte markers was. This may indicate incorrect lineage acquisition by the neural crest-derived stem cells. This set of studies also showed that the neural crest-derived stem cells migrated normally to the cornea but failed to express the appropriate markers indicating impaired differentiation rather than migration. The final set of studies were *in vitro* and showed that cells cultured from the periocular mesenchyme were responsive to TGFβ2 and subsequently expressed *Foxc1* and *Pitx2*. This corroborates the *in vivo* data. In conclusion these studies show the important role of *TGFβ2* in fate specification of neural crest-derived stem cells in the developing mouse eye.

Neural crest-derived stem cells from the mouse corneal stroma are multipotent

Yoshida *et al.* have shown that multipotent neural crest-derived stem cells can be isolated from the adult mouse corneal stroma (Yoshida *et al.*, 2006). They showed that clonal spheres could be obtained from single cells (which they termed corneal precursors) derived from the corneal stroma. As bone marrow-derived mesenchymal stem cells are multipotent and have been shown to contribute to non-haematopoietic lineages it was hypothesised that these clonal spheres could indeed be bone marrow-derived. Although they were CD34 positive, indicating possible bone marrow origin, they were not CD45 positive. Moreover, mice were transplanted with bone marrow cells from GFP positive mice and 8 weeks later corneal stromal clonal spheres were prepared from the transplanted mice. Although GFP positive cells were found within the cultures, these cells did not generate clonal spheres. This indicates that the corneal stromal clonal spheres are not bone marrow-derived. Due to the role of the neural crest in corneal stromal development, it was therefore hypothesised that the spheres were of neural crest origin. By generating spheres from transgenic mice, Yoshida *et al.* showed that indeed the clonal spheres generated from the adult corneal stroma were of neural crest origin. Moreover the spheres expressed markers of the embryonic neural crest, such as *Twist, Slug, Snail,* and *Sox9* confirming their neural crest origin. To show the multipotent nature of the stem cells from these clonal spheres, *in vitro* studies were conducted to show that they could differentiate into adipocytes, chondrocytes, and neuronal cells as well as keratocytes. Side population is a flow cytometric property based on Hoechst dye exclusion exhibited by many types of stem cells including haematopoietic stem cells, neuronal stem cells and indeed also by limbal stem cells. It was found in the experiments by Yoshida *et al.* that the clonal spheres were rich in side population cells. In conclusion, these studies show that the adult mouse corneal stroma consists of corneal precursors which are of neural crest origin and multipotent in nature.

Chromosomal aberration of subcultured neural crest-derived stem cells

Brandl *et al.* also conducted some further work looking at later passages of neural crest-derived stem cell-like cells from the limbus of juvenile mice (Brandl *et al.*, 2010). They reported that proliferation of these stem cells could be extended beyond 10 to 60 passages with a greater proliferative potential as highlighted by increased expression of the cell proliferation marker *Ki67* and increased number colony forming units. There was little change reported in the clonal size or morphology in the same cell line. The early passages from 3 to 15 again showed evidence of progenitor cell markers including *Abcg2* and *Nestin* and neural crest markers *Twist* and *Snail* supporting the neural crest stem cell properties of these cells. In addition it was reported that these limbal tissue-derived stem cells could be differentiated into adipocytes and osteoblasts. During later passages in particular there was a decline in the tumour suppressor gene *p21*. Karyotyping yielded normal structure and number up to passage 8 but there was evidence of chromosomal aberrations to at least tetraploidy during later subcultures. The *p21* and karyoptyping data shows that certainly during later stages of culturing these neural crest-derived stem cells there is a tendency for these cells to become neoplastic.

Conclusion

The identification, analysis and characterisation of neural crest-derived stem cells from the corneo-limbal stroma is very much in its infancy but there are some studies which highlight the important potential of these stem cells. This chapter highlights the experimental studies, which have taken place in this field to date, including quail-chick chimeric grafts, TGFβ2 related signalling mechanisms and *in vitro* characterisation and differentiation of these neural crest stem cells. The limbus has already provided us with the growing field of corneal epithelial stem cell biology which has led to advances in management of chemical eye burns (Kolli *et al.*, 2010; Shortt *et al.*, 2007). Although neural crest-derived stem cells from the corneo-limbal stroma have important differentiation potential, it remains to be elucidated whether this can be harnessed for therapeutic approaches.

References

Ahmad S, Figueiredo F, Lako M (2006). Corneal epithelial stem cells: Characterization, culture and transplantation. *Regen. Med.* 1(1): 29–44.

Ahmad S *et al.* (2010a). Stem cell therapies for ocular surface disease. *Drug Discov. Today* 15(7–8): 306–313.

Ahmad S *et al.* (2010b). The culture and transplantation of human limbal stem cells. *J. Cell Physiol.* 225(1): 15–19.

Brandl C *et al.* (2009). Identification of neural crest-derived stem cell-like cells from the corneal limbus of juvenile mice. *Exp. Eye Res.* 89(2): 209–217.

Brandl C *et al.* (2010). Spontaneous immortalization of neural crest-derived corneal progenitor cells after chromosomal aberration. *Cell Prolif.* 43(4): 372–377.

Hassell JR, Birk DE (2010). The molecular basis of corneal transparency. *Exp. Eye Res.* 91(3): 326–335.

Ittner LM *et al.* (2005). Compound developmental eye disorders following inactivation of TGF beta signalling in neural-crest stem cells. *J. Biol.* 4(3): 11.

Kolli S *et al.* (2010). Successful clinical implementation of corneal epithelial stem cell therapy for treatment of unilateral limbal stem cell deficiency. *Stem Cells* 28(3): 597–610.

Lwigale PY (2001). Embryonic origin of avian corneal sensory nerves. *Dev. Biol.* 239(2): 323–337.

Lwigale PY, Cressy PA, Bronner-Fraser M (2005). Corneal keratocytes retain neural crest progenitor cell properties. *Dev. Biol.* 288(1): 284–293.

McGowan SL *et al.* (2007). Stem cell markers in the human posterior limbus and corneal endothelium of unwounded and wounded corneas. *Mol. Vis.* 13: 1984–2000.

Shortt AJ *et al.* (2007). Transplantation of ex vivo cultured limbal epithelial stem cells: A review of techniques and clinical results. *Surv. Ophthalmol.* 52(5): 483–502.

Yoshida S *et al.* (2006). Isolation of multipotent neural crest-derived stem cells from the adult mouse cornea. *Stem Cells* 24(12): 2714–2722.

Zieske JD (2004). Corneal development associated with eyelid opening. *Int. J. Dev. Biol.* 48(8–9): 903–911.

PART II

ADULT NEURAL CREST STEM CELLS

4

CHARACTERISATION OF NEURAL CREST-DERIVED STEM CELLS IN DIFFERENT TISSUES

Narihito Nagoshi[1–4] and Hideyuki Okano[1]

[1]Department of Physiology, and [2]Department of Orthopedic Surgery, Keio University
[3]School of Medicine, 35 Shinanomachi, Shinjuku-ku, Tokyo 160-8582
[4]Clinical Research Centre, National Hospital Organization, Murayama Medical Centre, 2-37-1 Gakuen, Musashimurayama, Tokyo 208-0011, Japan

A burst of recent findings has shown that neural crest-derived stem cells (NCSCs) reside in diverse mammalian tissues. In addition to tissues known to be derived from the neural crest, recent studies show that NCSCs exist in tissues that are not neural crest-derived, such as bone marrow. Although how NCSCs are defined has varied among reports to date, it is clear that NCSCs can self-renew and have the potential to differentiate into several different neural-crest lineages, including neurons, glial cells, myofibroblasts, melanocytes, adipocytes, chondrocytes, osteocytes, and connective tissues (Crane & Trainor, 2006; Delfino-Machin *et al.*, 2007). NCSCs can express a wide range of characteristics, with their specific properties mainly depending on their tissue source and the animal's ontogenic stage. The identification of NCSCs in this wide variety of tissues opens an entirely new approach for developing autologous cell replacement therapies for use in regenerative medicine.

Characterisation of embryonic NCSCs

Stemple and Anderson were the first to describe mammalian NCSCs, which they isolated from the rat embryonic neural tube (Stemple & Anderson, 1992). These NCSCs were isolated using flow cytometry set to select cells expressing low affinity nerve growth factor (NGF) receptor (p75^{LNTR}). The frequency of colony formation was significantly higher in the p75$^+$ than in the p75$^-$ fraction. The p75$^+$ colony-forming cells were self-renewing and gave rise to neurons, glial cells, and myofibroblasts. These colony-forming cells are now well accepted to be NCSCs.

Cells with properties similar to NCSCs' have been isolated from the rat embryonic sciatic nerve in the post-migratory phase of neural crest development (Morrison *et al.*, 1999). Because glial cells in the sciatic nerve also express p75, the authors selected the cell fraction that was both positive for p75 and negative for P0 (peripheral myelin protein). The isolated P0$^-$p75$^+$ NCSCs were significantly enriched among the colony-forming cells, were self-renewing and multipotent. An *in vivo* study with mice revealed that NCSCs in the peripheral nerves generate Schwann cells and endoneurial fibroblasts during mouse embryonic development (Joseph *et al.*, 2004). These reports demonstrated that multipotent NCSCs persist at least until late gestation, after the onset of neural crest migration, and suggested that NCSCs persist in other tissues as well, during embryogenesis.

Rat NCSCs were also isolated from the gut at E14.5 by selecting for the p75$^+$ and α4 integrin$^+$ fraction (Bixby *et al.*, 2002). The authors compared the characteristics of the NCSCs from the gut with those from the sciatic nerve, and found that the gut NCSCs are sensitive to neurogenic factors, while sciatic nerve NCSCs are sensitive to gliogenic factors *in vitro*. Upon the transplantation into developing peripheral nerves *in vivo*, the gut NCSCs gave rise primarily to neurons, but the sciatic nerve NCSCs gave rise to glial cells. These results suggested that the phenotype of NCSCs mainly depends on their post-migratory tissue source. Furthermore, NCSCs express combinations of characteristics that reflect regional environmental differences as well as cell-intrinsic ones (Bixby *et al.*, 2002). However, the mechanisms that control the lineage determination and/or timing of differentiation among tissue sources remain unclear. In this respect, it will be interesting to clarify the molecular mechanisms affecting the epigenetic modifications of differentiation-related genes that regulate the spatial and temporal specifications of NCSCs.

Bone marrow

We demonstrated the existence of NCSCs in the bone marrow (BM) of adult rodents using Cre-lox system-mediated lineage analysis and the sphere culture technique (Nagoshi *et al.*, 2008). Transgenic mice expressing Cre recombinase under control of the P0 promoter (P0-Cre) (Yamauchi *et al.*, 1999) or the Wnt1 promoter/enhancer (Wnt1-Cre) (Danielian *et al.*, 1998) were crossed with EGFP reporter mice (CAG-CAT$^{loxP/lox}$P-EGFP) (Kawamoto *et al.*, 2000) to obtain P0-Cre/CAG-EGFP or Wnt1-Cre/CAG-EGFP double-transgenic mice (Nagoshi *et al.*, 2008). The P0 promoter expresses genes in neural crest-derived cells that have differentiated from the neural tube (Yamauchi *et al.*, 1999), and the Wnt1 promoter/enhancer expresses genes in the dorsal neural tube and roof-plate, from the onset of neural crest migration (Danielian *et al.*, 1998). In these transgenic mice, the transient activation of the P0 promoter and Wnt1 promoter/enhancer induces Cre-mediated recombination in pre-migratory neural crest cells, indelibly tagging them and their progenies with EGFP expression. In these mice, we observed EGFP$^+$ NCSCs migrating into the aorta-gonad-mesonephros (AGM) region during embryogenesis (Figure 1) (Nagoshi *et al.*, 2008). The AGM region is a transient embryonic tissue in which the first adult-type haematopoietic stem cells (HSCs) are generated. Late in embryogenesis, HSCs migrate through the bloodstream to the fetal liver; they enter the bone marrow (BM) just before birth (Dzierzak & Speck, 2008). We found that like the HSCs, NCSCs migrate from the AGM region through the bloodstream, through the fetal liver, and finally to the BM (Nagoshi *et al.*, 2008). The timing of the NCSC migration coincides with that of the HSCs, implying that some undefined relationship exists between the two stem cell types.

Prospectively isolated EGFP$^+$ cells from the BM of P0- and Wnt1-Cre/CAG-EGFP adult mice proliferate *in vitro* to form clonal spheres, are self-renewing, and differentiate into neurons, glial cells, and myofibroblasts (Nagoshi *et al.*, 2008). The presence of NCSCs in the BM was supported by a report using the same P0-Cre/CAG-EGFP reporter mice that

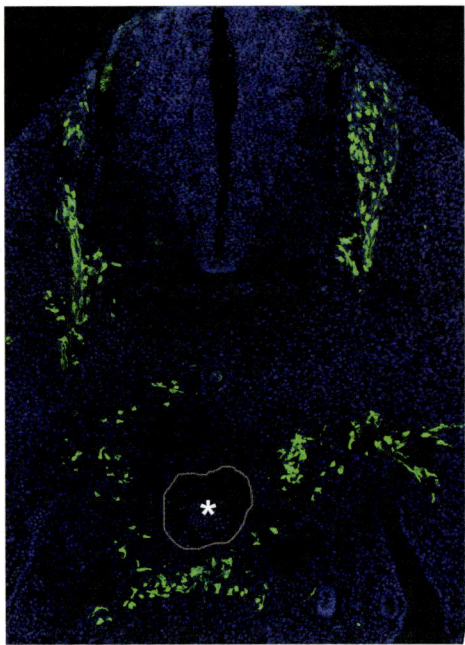

Figure 1. Migration pathway of neural crest cells to the BM in an embryonic mouse. In the transverse cross-section of an embryo, GFP+ neural crest cells migrate from the dorsal neural tube to the AGM region, which includes the aorta (asterisk), and then continue to migrate through the blood vessels to the BM.

demonstrated that a portion of the mesenchymal stem cells (MSCs) in the BM of the lower extremities are of neural crest lineage (Takashima *et al.*, 2007). That report also showed that the earliest lineage to generate MSCs during embryonic trunk development is the Sox1+ neuroepithelium, rather than the mesoderm, and at least some of these cells originated from the neural crest (Takashima *et al.*, 2007).

In addition, we recently showed that two putative markers, PDGFRα and Sca-1, can be used to isolate cells that are rich in MSCs. Some of the PDGFRα+Sca-1+ MSCs isolated from the BM of P0- and Wnt1-Cre/CAG-EGFP mice expressed EGFP, and could differentiate into adipocytes, chondrocytes, and osteocytes (Morikawa *et al.*, 2009a; Morikawa *et al.*, 2009b). These results supported the idea that neural crest-derived cells contribute to the MSC population. Considering that some of the MSCs are derived from the neural crest, NCSCs might be cellular components of the HSC niche, controlling the maintenance, proliferation, differentiation, and recruitment of HSCs in the BM (Uccelli *et al.*, 2008). This idea is supported by the findings that Nestin$^+$ MSCs in the rodent BM show a close physical association with HSCs, and have very high expression levels of core HSC maintenance genes (Mendez-Ferrer *et al.*, 2010), given that Nestin is also a marker for NCSCs (Nagoshi *et al.*, 2008; Tomita *et al.*, 2005; Yoshida *et al.*, 2006).

Several groups have demonstrated the presence of stem or precursor cells in the BM that can generate neurons. For example, bone marrow stromal cells (BMSCs) harvested from rat and human express Nestin and differentiate into neurons and glial cells *in vitro* (Sanchez-Ramos *et al.*, 2000). Another report demonstrated that rat and human BMSCs cultured with FBS expand as undifferentiated cells, and upon differentiation, become neurons (Woodbury *et al.*, 2000). The differentiation of BMSCs into functional neurons is enhanced by Noggin (Kohyama *et al.*, 2001). However, the developmental origin and differentiation potential of the BMSCs remain unclear, largely because it is difficult to understand how neurons, which are ectodermal in origin, can be generated by BM-derived cells. Our results demonstrating the migration pattern and presence of NCSCs in the BM suggest that this unusual differentiation potential could reflect the differentiation of NCSCs inhabiting the BM. It will be interesting to clarify the relationship between NCSCs and the BM-derived stem cells that are reported to generate neural cells.

Skin

One of the more intriguing findings about NCSCs is their existence in the skin. Cells isolated from juvenile and adult rodent skin proliferate to form spheres and differentiate into several types of cells: neurons, glial cells, smooth muscle cells, and adipocytes (Toma *et al.*, 2001). Importantly, a single cell can also form a sphere that shows self-renewal for at least 5 months of passaging and still shows multilineage differentiation into cells of both neural and mesodermal origins. These cells are called SKPs (Skin-derived precursor cells) (Toma *et al.*, 2001). Although the developmental origin of the SKPs was unclear when the report was published, the same group later demonstrated that the SKPs in facial skin are derived from the neural crest (Fernandes *et al.*, 2004). Using Wnt-Cre/ROSA-LacZ double-transgenic mice, a line used for neural-crest lineage tracing, they showed that whisker follicle dermal papillae are entirely neural crest–derived (Fernandes *et al.*, 2004). In addition, SKP-derived spheres from the facial skin of Wnt1-Cre/ROSA-LacZ mice are positive for β-galactosidase. The SKPs express the transcription factor genes *slug, snail, twist, Pax3*, and *Sox9*, which are also expressed in embryonic NCSCs (Fernandes *et al.*, 2004).

Recently, detailed analyses have shown that SKPs originate from $Sox2^+$ dermal precursors residing within the dermal sheath and dermal papilla of hair follicles and that these cells exhibit properties of dermal stem cells (Biernaskie *et al.*, 2009). Moreover, a recent report argues that SKPs are the likely cell of origin for the dermal tumors found in patients with neurofibrobromatosis type 1, a neural crest-related disorder in which patients develop malignant skin tumours composed of Schwann cells, melanocytes, and dermal fibroblasts, which further supports the similarity of SKPs to NCSCs (Le *et al.*, 2009). These reports collectively revealed that SKPs persist in the dermal papilla and dermal sheath.

In contrast, another group identified different NCSCs, "epidermal neural crest cells" (EPI-NCSCs), in the adult mouse whisker follicle, using a different approach (Sieber-Blum *et al.*, 2004). This group used Wnt1-Cre reporter mice to show marker expression in the bulge region of the follicle, which is a different location from that of the SKPs. Migrating EPI-NCSCs were observed in explants of the whisker follicle bulges. *In vitro* analysis of the emigrated EPI-NCSCs revealed that they can self-renew and differentiate into neurons, Schwann cells, smooth muscle cells, and melanocytes, highlighting the pluripotency of individual clones (Sieber-Blum *et al.*, 2004). These observations were corroborated by another report, in which a population of multipotent $Nestin^+K15^-$ cells was identified in the bulge region of rodent hair follicles, which could generate neural cells, keratinocytes, adipocytes, and smooth muscle cells *in vitro* and upon transplantation *in vivo* (Amoh *et al.*, 2005). Examination of the gene expression profile of the EPI-NCSCs by longSAGE (long serial analysis of gene expression) revealed 19 genes expressed in common between the EPI-NCSCs and embryonic NCSCs (Hu *et al.*, 2006). Although EPI-NCSCs and the epidermal stem cells that generate keratinocytes share the bulge as their stem-cell niche, they are readily distinguishable by their gene-expression profiles. Interestingly, these authors also examined the EPI-NCSCs for the expression of SKP cell markers, but the EPI-NCSCs did not express any of them, showing that the EPI-NCSCs are very different from SKPs (Fernandes *et al.*, 2004; Hu *et al.*, 2006). While EPI-NCSCs reside in the bulge region, SKPs are in the dermal papilla and dermal sheath, suggesting that hair follicles possess two different populations of NCSCs.

In addition to the whisker follicle of the facial skin, a recent study showed that a subpopulation of sphere-initiating cells from the murine trunk skin is also of neural crest origin (Wong *et al.*, 2006). Spheres derived from the trunk skin contain cells that express the NCSC markers p75 and Sox10, self-renew for more than 20 passages, and can differentiate into neurons, glial cells, smooth muscle cells, chondrocytes, melanocytes, and adipocytes. Using Desert Hedgehog (Dhh)-Cre/ROSA-LacZ mice, which express Cre recombinase in the peripheral glial lineage, Wong *et al.* also found that *LacZ*-positive cells in the bulge region of the trunk skin were positive for p75 and Sox10, suggesting that NCSCs that could give rise to the glial lineage residing there (Wong *et al.*, 2006). Moreover, in Dct-Cre/ROSA-*LacZ* mice, which express Cre recombinase in melanocytes, *LacZ*-positive cells in the bulge region and hair follicle bulb were positive for p75 and Sox10, suggesting that NCSCs that can give rise to the melanocyte lineage also exist in the bulge region and bulb. When these authors prospectively isolated the enhanced yellow fluorescent protein (EYFP)-positive cells from the trunk skin of Dhh-Cre/ROSA-EYFP and Dct-Cre/ROSA-EYFP mice, the cells proliferated to form spheres, and the spheres contained cells positive for p75 and Sox10. The authors concluded that NCSCs or neural crest-derived progenitor cells that are restricted to the glial and melanocyte lineages exist in the trunk skin of adult mice (Wong *et al.*, 2006).

Stem cells from human skin have been identified as well. SKPs from the human scalp express Nestin by immunohistochemistry and differentiate into neurons (Toma *et al.*, 2001). The same group also showed that SKPs exist in juvenile human foreskin that are self-renewing for several passages, differentiate into neurons, glial cells, smooth muscle cells, and adipocytes, and express the neural crest-specific markers *Pax3*, S*nail*, and S*lug* by RT-PCR (Toma *et al.*, 2005). Similar results were observed in the adult human dermis, which contains sphere-initiating cells (Joannides *et al.*, 2004). These spheres express markers for neural stem/progenitor cells, including Nestin and Musashi1 (Okano *et al.*, 2002; Sakakibara *et al.*, 1996), and differentiate into neurons and smooth muscle cells. However, they do not express neural crest markers or differentiate into glial cells (Joannides *et al.*, 2004).

Similar to the studies using rodent skin, recent reports have focused on the niche of human neural crest-like stem cells. Hunt *et al.* micro-dissected the dermal papillae of facial hair skin, and showed that single dermal papilla cells grow as spheres in SKP proliferation medium, express Sox10 and Pax3, and differentiate into neural cells (Hunt *et al.*, 2008). Interestingly, sphere cultures established by the micro-dissection of the dermal papilla show more reliable generation and better expansion of the spheres than the sphere cultures of whole dissociated skin (Hunt *et al.*, 2008). Thus, the dermal papilla seems to harbour NCSCs.

In contrast, Yu *et al.* demonstrated the existence of stem cells with neural crest characteristics in the bulge region of the human hair follicle (Yu *et al.*, 2006; Yu *et al.*, 2010). They developed a unique culture system by removing the mesenchymal compartment of the follicle to eliminate any potential contribution of dermal precursors. By culturing the intact hair follicles in the presence of FGF-2 and embryonic stem cell medium, they coaxed the stem cells to proliferate *in situ*, within their niche in the bulge region, and the cells formed spheres that erupted from the bulge region (Yu *et al.*, 2010). The authors compared the gene expression profiles between their stem cells and mouse EPI-NCSCs, and found that they expressed 18 of the 19 genes in the EPI-NCSC signature. The clonal cultured bulge stem cells self-renew and are multipotent, differentiating into neuronal cells that survive in the mouse brain after transplantation. Taken together, these studies show that the human skin, like that of the rodent, harbours dermal and epithelial precursors with characteristics of NCSCs (Hunt *et al.*, 2008; Yu *et al.*, 2010).

Note, however, that Yu *et al.* tried to identify the stem cells in the human hair follicle *in vivo* by immunohistochemistry, but it was difficult, because there are no specific markers for human NCSCs (Yu *et al.*, 2010). Since the sphere-initiating cells from human skin were retrospectively identified in all of these reports (Table 1), it remains uncertain whether

Table 1.   Isolation of NCSCs from various tissues. A table in our review article (Nagoshi *et al.*, 2009) was modified for this table.

Authors	Animal	Age	Tissue	Isolation	Marker	Genotype
Stemple	Rat	E10.5	Neural tube	Prospective	p75+	W/T
Morrison	Rat	E14.5	Sciatic nerve	Prospective	p75+P0−	W/T
Bixby	Rat	E14.5	Gut	Prospective	p75+α4+	W/T
Kruger	Rat	Adult	Gut	Prospective	p75+	W/T
Hagedorn	Rat	E14	DRG	Retrospective	p75+	W/T
Hjerling-Leffler	Mouse	E11.5	DRG	Retrospective		W/T
Li	Rat	Adult	DRG	Retrospective		W/T
Nagoshi	Mouse	Adult	DRG, whisker pad, bone marrow	Prospective	EGFP	P0 and Wnt1-Cre/Floxed-EGFP
Toma	Mouse	Juvenile and adult	Skin(face and back)	Retrospective		W/T
Sieber-Blum	Mouse	Adult	Skin(bulge)	Retrospective		Wnt1-Cre/ROSA-LacZ
Amoh	Mouse	Adult	Skin(bulge)	Retrospective		Nestin-driven GFP
Wong	Mouse	Adult	Skin(back)	Retrospective		W/T
	Mouse	Adult	Skin(back)	Prospective	EYFP	Dhh and Dct-Cre/ROSA-EYFP
Tomita	Mouse	Juvenile and adult	Heart	Prospective	SP cells	W/T, P0-Cre/Floxed-EGFP
Yoshida	Mouse	Adult	Cornea	Retrospective		W/T, P0 and Wnt1-Cre/Floxed-EGFP
Pardal	Rat	Adult	Carotid body	Retrospective		W/T
	Mouse	Adult	Carotid body	Prospective	EGFP	GFAP promoter-EGFP
Widera	Rat	Adult	Palatum	Retrospective		W/T
Toma	Human	Adult	Skin(scalp)	Retrospective		
Toma	Human	Children	Foreskin	Retrospective		
Hunt	Human	Adult	Skin(face and scalp)	Retrospective		
Yu	Human	Adult	Skin(scalp)	Retrospective		

these cells are truly derived from the neural crest. Future studies will help identify novel neural crest markers to allow the prospective isolation of NCSCs and their selective enrichment from other sources.

Dorsal root ganglia

Neural crest cells were first discovered in chick embryos as the precursors of the spinal sensory ganglia, the dorsal root ganglia (DRG) (His, 1868). In recent years, detailed analyses of mammalian NCSCs in the DRG have been carried out. In one report, single cells dissociated from rat embryonic DRGs were labelled with p75 by live-cell staining, and the identified p75$^+$ NCSCs were shown to give rise to neurons, glial cells, and smooth muscle-like cells in response to instructive extracellular cues, but their self-renewal activity was not assessed (Hagedorn *et al.*, 1999). Another study showed that neural crest boundary cap cells, which are found in embryos, can generate neurons and satellite cells (Maro *et al.*, 2004). Boundary cap cells are neural crest derivatives that form clusters at the entry and exit points of peripheral nerve roots; they migrate to and colonise the DRG during embryogenesis. They behave as gatekeepers of the CNS–PNS boundary, because they prevent motor neuron somata from exiting toward the periphery, but they allow axons to grow through the exit points (Vermeren *et al.*, 2003).

Boundary cap clusters contain NCSCs that self-renew, show multipotency, and differentiate into mature sensory neurons and Schwann cells under appropriate conditions (Aquino *et al.*, 2006; Hjerling-Leffler *et al.*, 2005). These findings raised the possibility that NCSCs might persist in the DRGs throughout life, and this was later demonstrated (Li *et al.*, 2007). Interestingly, these NCSCs probably originate from satellite cells (Li *et al.*, 2007). Given that some of the satellite cells are thought to be derived from boundary cap cells (Maro *et al.*, 2004; Zirlinger *et al.*, 2002), these data indicate that the NCSCs form a subpopulation of boundary cap cells that migrate into the DRG during embryogenesis, where they are maintained in an undifferentiated state throughout the life of the animal.

Our group also confirmed the existence of NCSCs in the DRG of adult mice (Nagoshi *et al.*, 2008). In this study, we compared the characteristics of NCSCs in various tissues of adult mice by examining the expression levels of the NCSC markers *sox10* (Paratore *et al.*, 2001) and *p75*

(Stemple & Anderson, 1992) and of markers for neural stem/progenitor cells, *nestin* (Lendahl *et al.*, 1990) and *musashi1* (Okano *et al.*, 2002; Sakakibara *et al.*, 1996). We found that these markers were expressed at higher levels in NCSCs from the DRG than in NCSCs from the whisker pad or BM. These results may reflect the self-renewal activity and multipotency of NCSCs, because the DRG-derived NCSCs displayed a greater ability to form secondary spheres and a higher proportion of cells that maintained a multilineage differentiation potential (Nagoshi *et al.*, 2008). Although the methods for identifying NCSCs and the culture conditions were different in the various reports described above, the combined findings support the idea that the DRG contains a high proportion of NCSCs.

Other tissues

NCSCs have been identified in the adult rat gut as well as that of the embryo. A comparison of fetal and adult gut NCSCs showed that the adult NCSCs self-renew less efficiently and differentiate into a narrower range of neuronal subtypes (Bixby *et al.*, 2002; Kruger *et al.*, 2002). Considering that the characteristics of these NCSCs differ according to their spatial and temporal context (Bixby *et al.*, 2002; Kruger *et al.*, 2002; Nagoshi *et al.*, 2008), it is not possible to categorise these NCSCs as a homogenous population. It will be important to classify these populations according to their differentiation potential and self-renewal activity, and to elucidate the molecular mechanisms for their maintenance and lineage determination, with respect to their spatial and temporal contexts.

Another type of NCSC has been identified in the heart of adult mice (Tomita *et al.*, 2005). Cardiac side population (SP) cells contain a subpopulation of NCSCs that can generate spheres and differentiate into neurons, glial cells, smooth muscle cells, and cardiomyocytes. In immunohistochemical analyses using P0-Cre/CAG-EGFP adult heart tissue, Nestin-positive cells were identified among the EGFP-positive cells that proliferated to form spheres *in vitro*. These findings suggested that NCSCs that can differentiate into various cell types remain in the heart of adult mice.

NCSCs have been identified in the adult mouse cornea, as well (Yoshida *et al.*, 2006). Cornea-derived spheres express Nestin and Musashi1, self-renew over several passages, and differentiate into neural- and mesenchymal-lineage cells. The NCSCs in the cornea are also enriched in SP cells, like the cardiac NCSCs (Tomita *et al.*, 2005; Yoshida *et al.*, 2006). Cornea-derived cells from P0-Cre/CAG-EGFP and Wnt1-Cre/CAG-EGFP adult mice proliferate to form EGFP$^+$ spheres, indicating the existence of NCSCs in the adult cornea.

Another study demonstrated the existence of NCSCs in the carotid body, an oxygen-sensing organ of the sympathoadrenal lineage that grows under conditions of hypoxemia (Pardal *et al.*, 2007). GFAP$^+$ cells in the rat carotid body incorporate BrdU *in vivo*, and proliferate *in vitro* to form spheres that differentiate into tyrosine hydroxylase (TH)-positive neurons and smooth muscle cells, suggesting that the GFAP$^+$ cells are stem/progenitor cells that resemble NCSCs in some aspects. Although the GFAP$^+$ stem cells are reversibly converted to Nestin$^+$ progenitors in re-normoxia, the equilibrium is displaced toward the Nestin$^+$ progenitors, which give rise to TH$^+$ neurons under hypoxic conditions (Pardal *et al.*, 2007).

A recent report revealed the presence of Nestin$^+$ neural crest-related stem cells within the Meissner corpuscles and Merkel cell-neurite complexes located in the palatum of adult rodents (Widera *et al.*, 2009). The cells were dissociated from the palatum and cultured to make spheres, which were kept in culture for more than 20 passages without losing their chromosomal stability and capacities for proliferation and differentiation. These spheres have the capacity to differentiate into neurons and glial cells (Widera *et al.*, 2009). It would be interesting to trace genetic markers using P0 and Wnt1-Cre mice to examine the origin of these sphere-initiating cells.

Although NCSCs from various adult tissues have been reported, it would be rash to conclude that all tissue-derived stem cells are NCSCs. For example, multipotent precursors that generate neural and pancreatic lineages have been identified in the adult mouse pancreas (Seaberg *et al.*, 2004), and they do not express the neural crest markers *Pax3*, *Twist*, *Sox10*, or *Wnt1* by RT-PCR. The authors concluded that the precursors are not neural crest derivatives. On the other hand, these cells do express *slug*, *snail*, and *p75*, and therefore the possibility that they are derived from the neural crest cannot be excluded, especially because the expression patterns of neural crest markers in NCSCs are quite different, depending on the tissue source (Nagoshi *et al.*, 2008). Indeed, a recent study demonstrated that NCSCs migrate toward the pancreas and play an important role in beta cell differentiation by regulating the beta cell mass during development (Nekrep *et al.*, 2008), raising the possibility that NCSCs persist in the pancreas into adulthood.

Various methods for identifying NCSCs

So far, a single marker has not been identified for the isolation of NCSCs, although several research groups have established original methods for identifying NCSCs (Table 1). Comparison of these protocols indicates that prospective rather than retrospective isolation is much better for purifying cells in their native condition. Retrospective isolation raises the possibility of contamination by various non-NCSCs and increases the chance that the characteristics of the NCSCs will change during the cell culture procedure. In rodent studies, p75 has been found to be a good marker for the prospective isolation of NCSCs, and it has been widely used for NCSC purification by several groups (Bixby *et al.*, 2002; Kruger *et al.*, 2002; Morrison *et al.*, 1999). Considering that genetic lineage labelling techniques such as P0-Cre and/or Wnt1-Cre/CAG-EGFP are available for mice, one of the best ways to purify NCSCs to date has been to isolate p75$^+$ EGFP$^+$ cells by flow-cytometry. Although NCSC markers including p75 that allow prospective isolation have been identified in rodents, no such marker has been established for human NCSCs. The possible identification of specific surface antigens for human NCSCs needs to be pursued further.

Application of NCSCs to regenerative medicine

The NCSC is one of the most promising cells being considered for use in regenerative medicine, because it is easily harvested from accessible peripheral tissues, which could increase the feasibility of autologous transplantation. Autologous transplantation would avoid immunological complications as well as the ethical concerns associated with the use of embryonic stem cells. Moreover, to our knowledge, transplanted NCSCs show no tumorigenicity. Of the various NCSCs, the research on skin-derived NCSCs is the most advanced, because of their accessibility.

One of the critical questions for the application of NCSCs to regenerative medicine is whether the cells that are differentiated from NCSCs are functional. Some evidence supports this hypothesis. Cultured rodent and human SKPs generate Schwann cells when treated with neuregulins, and myelinate host axons after their transplantation to injured peripheral nerves (McKenzie *et al.*, 2006). These Schwann cells also myelinate axons in the CNS when transplanted into the brain. Furthermore, the SKP-derived Schwann cells were transplanted into the injured spinal cord of the rat, and improved the locomotor function (Biernaskie *et al.*, 2007). This was the first report that NCSC-derived cells could contribute to the recovery of function following central nervous system injury, but these SKP-derived Schwann cells were harvested from neonatal murine trunk skin, not adult skin (Biernaskie *et al.*, 2007).

Nonetheless, recent reports demonstrated that adult EPI-NCSCs promote the recovery of sensory function in a model of mouse spinal cord injury (Hu *et al.*, 2010; Sieber-Blum *et al.*, 2006). The transplantation of EPI-NCSCs harvested from the bulge region of whisker follicles in adult mice into the contused spinal cord, resulted in a 24% improvement in sensory connectivity and a substantial recovery of touch perception 4 months after injury. The subsets of grafted cells acquired properties of functional neurons, and the others differentiated into myelinating glia. Finally, Hu *et al.* (2010) provided evidence that EPI-NCSCs express multiple genes that encode neurotrophic factors, angiogenic factors, and metalloproteases, the activities of which could account for the functional improvements.

In addition to CNS repair, neurons and Schwann cells derived from NCSCs in the bulge region of adult rats contribute to the functional repair of sciatic nerves (Lin *et al.*, 2009). An acellular nerve xenograft seeded with NCSC-derived neurons and Schwann cells bridged 4 cm gaps in rat sciatic nerves, and a number of regenerated axons were observed, with chemical synaptic structures that included synapsin 1 (Lin *et al.*, 2009).

Boundary cap cells, neural crest derivatives in the DRG of embryos that contain NCSCs, can remyelinate demyelinated CNS axons (Zujovic *et al.*, 2010). When grafted away from the lesion in the demyelinated spinal cord of mice, the progeny of boundary cap cells, in contrast to committed

Schwann cells, show a high migratory potential mediated by enhanced interactions with astrocytes and white matter. In response to demyelinated axons of the CNS, the boundary cap progeny generate myelin-forming Schwann cells (Zujovic *et al.*, 2010). Although boundary cap cells are harvested not from adults but from embryos, this report indicates that NCSCs derived from the DRG may potentially be applied to regenerative medicine, especially for use in the CNS.

A recent report demonstrated that NCSCs contribute to the repair of skeletal injury. Rat SKPs, which can differentiate into osteogenic and chondrogenic lineages, were transplanted into a tibial bone fracture model of NOD-SCID mice, and some of the transplanted cells adopted a mature osteocyte phenotype and integrated into the newly formed bone (Lavoie *et al.*, 2009). Moreover, some transplanted cells also differentiated into chondrocytes and into smooth muscle cells and pericytes that were associated with blood vessels. Thus, the micro-environment within the injured bone is sufficient to instruct the SKPs to differentiate along an osteogenic lineage, as with endogenous mesenchymal precursors (Lavoie *et al.*, 2009).

Future prospects

Recent reports have revealed the existence of NCSCs in various mammalian tissues. Surprisingly, NCSCs were detected in the BM and in skin beyond the expected germ layer (Fernandes *et al.*, 2004; Nagoshi *et al.*, 2008; Sieber-Blum *et al.*, 2004). These results raise the possibility that NCSCs might be found in other unexpected tissue sources. In addition to identifying NCSCs, it will be even more interesting to elucidate the physiological roles of NCSCs. It is generally thought that the stem cells from various adult tissues retain the capacity for tissue repair. Therefore, it is also likely that NCSCs have undiscovered biological and pathological roles that may be extremely helpful in the treatment of human disease. The currently known properties of adult NCSCs already make them attractive candidates for regenerative therapies such as the cell-replacement therapy. NCSCs can be harvested from autologous tissue sources, avoiding ethical and immunological problems. Future challenges include the establishment of culture systems that support the effective proliferation of adult

NCSCs, similar to those already established for ES cells and iPS cells, and the efficient differentiation of NCSCs into cells that are the most appropriate for treating specific lesions.

References

Amoh Y, Li L, Katsuoka K, Penman S, Hoffman RM (2005). Multipotent nestin-positive, keratin-negative hair-follicle bulge stem cells can form neurons. *Proc. Natl. Acad. Sci. USA* 102(15): 5530–5534.

Aquino JB, Hjerling-Leffler J, Koltzenburg M, Edlund T, Villar MJ, Ernfors P (2006). *In vitro* and *in vivo* differentiation of boundary cap neural crest stem cells into mature Schwann cells. *Exp. Neurol.* 198(2): 438–449.

Biernaskie J, Paris M, Morozova O, Fagan BM, Marra M, Pevny L, Miller FD (2009). SKPs derive from hair follicle precursors and exhibit properties of adult dermal stem cells. *Cell Stem Cell* 5(6): 610–623.

Biernaskie J, Sparling JS, Liu J, Shannon CP, Plemel JR, Xie Y, Miller FD, Tetzlaff W (2007). Skin-derived precursors generate myelinating Schwann cells that promote remyelination and functional recovery after contusion spinal cord injury. *J. Neurosci.* 27(36): 9545–9559.

Bixby S, Kruger GM, Mosher JT, Joseph NM, Morrison SJ (2002). Cell-intrinsic differences between stem cells from different regions of the peripheral nervous system regulate the generation of neural diversity. *Neuron* 35(4): 643–656.

Crane JF, Trainor PA (2006). Neural crest stem and progenitor cells. *Annu. Rev. Cell Dev. Biol.* 22: 267–286.

Danielian PS, Muccino D, Rowitch DH, Michael SK, McMahon AP (1998). Modification of gene activity in mouse embryos in utero by a tamoxifen-inducible form of Cre recombinase. *Curr. Biol.* 8(24): 1323–1326.

Delfino-Machin M, Chipperfield TR, Rodrigues FS, Kelsh RN (2007). The proliferating field of neural crest stem cells. *Dev. Dyn.* 236(12): 3242–3254.

Dzierzak E, Speck NA (2008). Of lineage and legacy: The development of mammalian hematopoietic stem cells. *Nat. Immunol.* 9(2): 129–136.

Fernandes KJ, McKenzie IA, Mill P, Smith KM, Akhavan M, Barnabe-Heider F, Biernaskie J, Junek A, Kobayashi NR, Toma JG and others (2004). A dermal niche for multipotent adult skin-derived precursor cells. *Nat. Cell Biol.* 6(11): 1082–1093.

Hagedorn L, Suter U, Sommer L (1999). P0 and PMP22 mark a multipotent neural crest-derived cell type that displays community effects in response to TGF-beta family factors. *Development* 126(17): 3781–3794.

His W (1868). Untersuchungen uber die erste Analge des Wirbeltierleibes. Die erste Entwicklung des Hunchens im Ei Leipzig 16: 237.

Hjerling-Leffler J, Marmigere F, Heglind M, Cederberg A, Koltzenburg M, Enerback S, Ernfors P (2005). The boundary cap: A source of neural crest stem cells that generate multiple sensory neuron subtypes. *Development* 132(11): 2623–2632.

Hu YF, Gourab K, Wells C, Clewes O, Schmit BD, Sieber-Blum M (2010). Epidermal neural crest stem cell (EPI-NCSC)-mediated recovery of sensory function in a mouse model of spinal cord injury. *Stem Cell Rev.* 6(2): 186–198.

Hu YF, Zhang ZJ, Sieber-Blum M (2006). An epidermal neural crest stem cell (EPI-NCSC) molecular signature. *Stem Cells* 24(12): 2692–2702.

Hunt DP, Morris PN, Sterling J, Anderson JA, Joannides A, Jahoda C, Compston A, Chandran S (2008). A highly enriched niche of precursor cells with neuronal and glial potential within the hair follicle dermal papilla of adult skin. *Stem Cells* 26(1): 163–172.

Joannides A, Gaughwin P, Schwiening C, Majed H, Sterling J, Compston A, Chandran S (2004). Efficient generation of neural precursors from adult human skin: Astrocytes promote neurogenesis from skin-derived stem cells. *Lancet* 364(9429): 172–178.

Joseph NM, Mukouyama YS, Mosher JT, Jaegle M, Crone SA, Dormand EL, Lee KF, Meijer D, Anderson DJ, Morrison SJ (2004). Neural crest stem cells undergo multilineage differentiation in developing peripheral nerves to generate endoneurial fibroblasts in addition to Schwann cells. *Development* 131(22): 5599–5612.

Kawamoto S, Niwa H, Tashiro F, Sano S, Kondoh G, Takeda J, Tabayashi K, Miyazaki J (2000). A novel reporter mouse strain that expresses enhanced green fluorescent protein upon Cre-mediated recombination. *FEBS Lett.* 470(3): 263–268.

Kohyama J, Abe H, Shimazaki T, Koizumi A, Nakashima K, Gojo S, Taga T, Okano H, Hata J, Umezawa A (2001). Brain from bone: Efficient "meta-differentiation" of marrow stroma-derived mature osteoblasts to neurons with Noggin or a demethylating agent. *Differentiation* 68(4–5): 235–244.

Kruger GM, Mosher JT, Bixby S, Joseph N, Iwashita T, Morrison SJ (2002). Neural crest stem cells persist in the adult gut but undergo changes in self-renewal, neuronal subtype potential, and factor responsiveness. *Neuron* 35(4): 657–669.

Lavoie JF, Biernaskie JA, Chen Y, Bagli D, Alman B, Kaplan DR, Miller FD (2009). Skin-derived precursors differentiate into skeletogenic cell types and contribute to bone repair. *Stem Cells Dev.* 18(6): 893–906.

Le LQ, Shipman T, Burns DK, Parada LF (2009). Cell of origin and microenvironment contribution for NF1-associated dermal neurofibromas. *Cell Stem Cell* 4(5): 453–463.

Lendahl U, Zimmerman LB, McKay RD (1990). CNS stem cells express a new class of intermediate filament protein. *Cell* 60(4): 585–595.

Li HY, Say EH, Zhou XF (2007). Isolation and characterization of neural crest progenitors from adult dorsal root ganglia. *Stem Cells* 25(8): 2053–2065.

Lin H, Liu F, Zhang C, Zhang Z, Guo J, Ren C, Kong Z (2009). Pluripotent hair follicle neural crest stem-cell-derived neurons and schwann cells functionally repair sciatic nerves in rats. *Mol. Neurobiol.* 40(3): 216–223.

Maro GS, Vermeren M, Voiculescu O, Melton L, Cohen J, Charnay P, Topilko P (2004). Neural crest boundary cap cells constitute a source of neuronal and glial cells of the PNS. *Nat. Neurosci.* 7(9): 930–938.

McKenzie IA, Biernaskie J, Toma JG, Midha R, Miller FD (2006). Skin-derived precursors generate myelinating Schwann cells for the injured and dysmyelinated nervous system. *J. Neurosci.* 26(24): 6651–6660.

Mendez-Ferrer S, Michurina TV, Ferraro F, Mazloom AR, Macarthur BD, Lira SA, Scadden DT, Ma'ayan A, Enikolopov GN, Frenette PS (2010). Mesenchymal and haematopoietic stem cells form a unique bone marrow niche. *Nature* 466(7308): 829–834.

Morikawa S, Mabuchi Y, Kubota Y, Nagai Y, Niibe K, Hiratsu E, Suzuki S, Miyauchi-Hara C, Nagoshi N, Sunabori T and others (2009a). Prospective identification, isolation, and systemic transplantation of multipotent mesenchymal stem cells in murine bone marrow. *J. Exp. Med.* 206(11): 2483–2496.

Morikawa S, Mabuchi Y, Niibe K, Suzuki S, Nagoshi N, Sunabori T, Shimmura S, Nagai Y, Nakagawa T, Okano H and others (2009b). Development of mesenchymal stem cells partially originate from the neural crest. Biochem. *Biophys. Res. Commun.* 379(4): 1114–1119.

Morrison SJ, White PM, Zock C, Anderson DJ (1999). Prospective identification, isolation by flow cytometry, and *in vivo* self-renewal of multipotent mammalian neural crest stem cells. *Cell* 96(5): 737–749.

Nagoshi N, Shibata S, Kubota Y, Nakamura M, Nagai Y, Satoh E, Morikawa S, Okada Y, Mabuchi Y, Katoh H and others (2008). Ontogeny and multipotency of neural crest-derived stem cells in mouse bone marrow, dorsal root ganglia, and whisker pad. *Cell Stem Cell* 2(4): 392–403.

Nagoshi N, Shibata S, Nakamura M, Matsuzaki Y, Toyama Y, Okano H (2009). Neural crest-derived stem cells display a wide variety of characteristics. *J. Cell Biochem.* 107(6): 1046–1052.

Nekrep N, Wang J, Miyatsuka T, German MS (2008). Signals from the neural crest regulate beta-cell mass in the pancreas. *Development* 135(12): 2151–2160.

Okano H, Imai T, Okabe M (2002). Musashi: A translational regulator of cell fate. *J. Cell Sci.* 115(Pt 7): 1355–1359.

Paratore C, Goerich DE, Suter U, Wegner M, Sommer L (2001). Survival and glial fate acquisition of neural crest cells are regulated by an interplay between the transcription factor Sox10 and extrinsic combinatorial signalling. *Development* 128(20): 3949–3961.

Pardal R, Ortega-Saenz P, Duran R, Lopez-Barneo J (2007). Glia-like stem cells sustain physiologic neurogenesis in the adult mammalian carotid body. *Cell* 131(2): 364–377.

Sakakibara S, Imai T, Hamaguchi K, Okabe M, Aruga J, Nakajima K, Yasutomi D, Nagata T, Kurihara Y, Uesugi S and others (1996). Mouse-Musashi-1, a neural RNA-binding protein highly enriched in the mammalian CNS stem cell. *Dev. Biol.* 176(2): 230–242.

Sanchez-Ramos J, Song S, Cardozo-Pelaez F, Hazzi C, Stedeford T, Willing A, Freeman TB, Saporta S, Janssen W, Patel N and others (2000). Adult bone marrow stromal cells differentiate into neural cells *in vitro*. *Exp. Neurol.* 164(2): 247–256.

Seaberg RM, Smukler SR, Kieffer TJ, Enikolopov G, Asghar Z, Wheeler MB, Korbutt G, van der Kooy D (2004). Clonal identification of multipotent precursors from adult mouse pancreas that generate neural and pancreatic lineages. *Nat. Biotechnol.* 22(9): 1115–1124.

Sieber-Blum M, Grim M, Hu YF, Szeder V (2004). Pluripotent neural crest stem cells in the adult hair follicle. *Dev. Dyn.* 231(2): 258–269.

Sieber-Blum M, Schnell L, Grim M, Hu YF, Schneider R, Schwab ME (2006). Characterization of epidermal neural crest stem cell (EPI-NCSC) grafts in the lesioned spinal cord. Mol. *Cell Neurosci.* 32(1–2): 67–81.

Stemple DL, Anderson DJ (1992). Isolation of a stem cell for neurons and glia from the mammalian neural crest. *Cell* 71(6): 973–985.

Takashima Y, Era T, Nakao K, Kondo S, Kasuga M, Smith AG, Nishikawa S (2007). Neuroepithelial cells supply an initial transient wave of MSC differentiation. *Cell* 129(7): 1377–1388.

Toma JG, Akhavan M, Fernandes KJ, Barnabe-Heider F, Sadikot A, Kaplan DR, Miller FD (2001). Isolation of multipotent adult stem cells from the dermis of mammalian skin. *Nat. Cell Biol.* 3(9): 778–784.

Toma JG, McKenzie IA, Bagli D, Miller FD (2005). Isolation and characterization of multipotent skin-derived precursors from human skin. *Stem Cells* 23(6): 727–737.

Tomita Y, Matsumura K, Wakamatsu Y, Matsuzaki Y, Shibuya I, Kawaguchi H, Ieda M, Kanakubo S, Shimazaki T, Ogawa S and others (2005). Cardiac neural crest cells contribute to the dormant multipotent stem cell in the mammalian heart. *J. Cell Biol.* 170(7): 1135–1146.

Uccelli A, Moretta L, Pistoia V (2008). Mesenchymal stem cells in health and disease. *Nat. Rev. Immunol.*

Vermeren M, Maro GS, Bron R, McGonnell IM, Charnay P, Topilko P, Cohen J (2003). Integrity of developing spinal motor columns is regulated by neural crest derivatives at motor exit points. *Neuron.* 37(3): 403–415.

Widera D, Zander C, Heidbreder M, Kasperek Y, Noll T, Seitz O, Saldamli B, Sudhoff H, Sader R, Kaltschmidt C and others (2009). Adult palatum as a novel source of neural crest-related stem cells. *Stem Cells* 27(8): 1899–1910.

Wong CE, Paratore C, Dours-Zimmermann MT, Rochat A, Pietri T, Suter U, Zimmermann DR, Dufour S, Thiery JP, Meijer D and others (2006). Neural crest-derived cells with stem cell features can be traced back to multiple lineages in the adult skin. *J. Cell Biol.* 175(6): 1005–1015.

Woodbury D, Schwarz EJ, Prockop DJ, Black IB (2000). Adult rat and human bone marrow stromal cells differentiate into neurons. *J. Neurosci. Res.* 61(4): 364–370.

Yamauchi Y, Abe K, Mantani A, Hitoshi Y, Suzuki M, Osuzu F, Kuratani S, Yamamura K (1999). A novel transgenic technique that allows specific marking of the neural crest cell lineage in mice. *Dev. Biol.* 212(1): 191–203.

Yoshida S, Shimmura S, Nagoshi N, Fukuda K, Matsuzaki Y, Okano H, Tsubota K (2006). Isolation of multipotent neural crest-derived stem cells from the adult mouse cornea. *Stem Cells* 24(12): 2714–2722.

Yu H, Fang D, Kumar SM, Li L, Nguyen TK, Acs G, Herlyn M, Xu X (2006). Isolation of a novel population of multipotent adult stem cells from human hair follicles. *Am. J. Pathol.* 168(6): 1879–1888.

Yu H, Kumar SM, Kossenkov AV, Showe L, Xu X (2010). Stem cells with neural crest characteristics derived from the bulge region of cultured human hair follicles. *J. Invest. Dermatol.* 130(5): 1227–1236.

Zirlinger M, Lo L, McMahon J, McMahon AP, Anderson DJ (2002). Transient expression of the bHLH factor neurogenin-2 marks a subpopulation of neural crest cells biased for a sensory but not a neuronal fate. *Proc. Natl. Acad. Sci. USA* 99(12): 8084–8089.

Zujovic V, Thibaud J, Bachelin C, Vidal M, Coulpier F, Charnay P, Topilko P, Baron-Van Evercooren A (2010). Boundary cap cells are highly competitive for CNS remyelination: Fast migration and efficient differentiation in PNS and CNS myelin-forming cells. *Stem Cells* 28(3): 470–479.

5

EPIDERMAL NEURAL CREST STEM CELLS

Oliver Clewes and Maya Sieber-Blum

Introduction

As described elsewhere in this volume, the neural crest is a transient embryonic tissue formed during neurulation. Following an epithelial to mesenchymal transition, neural crest cells delaminate from the overlying neural folds. Neural crest cells are highly migratory and translocate to numerous terminal destinations where they contribute to a wide array of cell types and tissues. Neural crest progeny include the autonomic nervous system, the enteric nervous system, most primary sensory neurons, tooth papillae, craniofacial bone/cartilage; they contribute to the septation of the cardiac outflow tract and contribute to its musculature and that of the great vessels. Neural crest cells also give rise to all pigment cells (melanocytes) of the skin and internal organs (Le Douarin & Kalcheim, 1999; Jiang *et al.*, 2002; Sieber-Blum, 2004). Due to this ability to form numerous different cell types, neural crest stem cells are of particular interest. Furthermore, the discovery that neural crest-derived stem cells persist in various postnatal locations makes them attractive candidates for clinical applications. The bulge of hair follicles contains neural crest-derived stem cells (Sieber-Blum *et al.*, 2004; Sieber-Blum & Grim, 2004). Using Wnt1-Cre:R26R compound transgenic mice, in which all neural crest cells and their descendents are specifically and indelibly marked by expression of β-galactosidase and appear blue upon Xgal reaction, we have shown that the bulge of hair follicles is a niche for neural crest-derived stem cells (Sieber-Blum *et al.*, 2004; Sieber-Blum & Grim, 2004). *In vitro* clonal analyses have shown that they are multipotent stem cells able to undergo self-renewal and to generate all major neural crest derivatives (Sieber-Blum *et al.*, 2004; Sieber-Blum & Grim, 2004). The follicular bulge thus contains at least two types of stem cells, epidermal stem cells and neural crest-derived stem cells. Due to their location in the epidermal bulge, the hair follicle-derived cells were termed epidermal-neural crest stem cells (EPI-NCSC). Taken together, the bulge of hair follicles provides a readily accessible location of EPI-NCSC that can be isolated by minimal invasive procedures. Due to their migratory ability, EPI-NCSC can be obtained as a highly pure population of stem cells. EPI-NCSC are thus attractive candidates for future cell replacement therapies and regenerative medicine.

Characterisation of mouse EPI-NCSC

Mouse EPI-NCSC are multipotent stem cells able to undergo self-renewal and to generate all major neural crest derivatives, including neurons, Schwann cells, smooth muscle cells, bone/cartilage cells, and pigment cells (Sieber-Blum *et al.*, 2004).

Three transcriptomes were generated by long serial analysis of gene expression (longSAGE); one from RNA of mouse EPI-NCSC, a second one from embryonic neural crest cells derived from E9.5 neural tube explants, and a third one from *in vitro* differentiated neural crest progeny. Three-way *in silico* comparisons, and a comparison with an epidermal stem cell gene expression profile, resulted in a panel of 19 genes that are expressed in bulge derived-neural crest cells and embryonic neural crest cells but not in epidermal stem cells, with which EPI-NCSC share the stem cell niche in the bulge. This panel of genes is termed a neural crest stem cell molecular signature (Hu *et al.*, 2006). Comparison analyses with data from the existing literature showed the EPI-NCSC are unique among known skin-resident stem cells. In particular, neural crest-derived stem cells differ from skin-derived precursors (SKPs), which are present in the dermal papilla and other locations of hairy and glabrous skin (Toma *et al.*, 2001; Toma *et al.*, 2005; Fernandes *et al.*, 2008; Hunt *et al.*, 2008).

The neural crest stem cell molecular signature consists of genes with a wide range of functions essential for maintaining stemness, regulation of lineage decisions, and metabolic functions. Some of the signature genes are well known neural crest disease genes (Hu *et al.*, 2006). Vars2, H1fx, Thop1, Ube4b, Adam12, and *Calr* have roles in the maintenance of protein metabolism and other functions. Vars2 is also involved in RNA metabolism. Targeted mutations in calretinin (*Calr*) cause neural tube defects, increased apoptosis in cardiomyocytes, and mid- to late-gestational lethality. Homozygous Adam12 mouse mutants die postnatally, and mutations in the human ADAM12 gene cause cardiomyopathies. Msx2 is a master regulatory gene in neurogenesis, osteogenesis and myogenesis. Loss of Msx2 function causes craniofacial defects, whereas Msx2 over-expression causes aortico-pulmonary septation defects. Msx2 mouse mutants also have alopecia. Pygo2 is a member of Wnt signalling pathways, which are essential for neural crest development and migration. Vdac1 is an intracellular transporter. Cryab is involved in developmental control and the Ets1 transcription factor is important for control of the cell cycle. Myo10 is required for the cell cytoskeleton and is involved in migration. Overall, the signature genes have functions appropriate of neural crest cells.

Mouse EPI-NCSC (mEPI-NCSC) grafts cause functional improvement in mouse models of spinal cord injury

The significance and potential of mEPI-NCSC has begun to be realised in animal models of human disease. In mouse models of spinal cord injury, grafted mEPI-NCSC caused an improvement in sensory connectivity and touch perception. Following grafting, mEPI-NCSC integrates with surrounding host spinal tissue. Importantly they did not migrate from the site of grafting and they did not form tumours (Sieber-Blum *et al.*, 2006), an essential prerequisite for stem cell-based therapies. Subsets of intraspinal mEPI-NCSC expressed markers for neurons, including GABAergic neurons, or markers for oligodendrocytes (Sieber-Blum *et al.*, 2006). Immuno-electronmicroscopy showed that some grafted mEPI-NCSC differentiated into myelinating glia. Interestingly, none of the grafted cells expressed glial fibrillary acidic protein (GFAP), which is a maker for Schwann cells and astrocytes. Spinal cord sensory evoked potentials showed a 24%

improvement in sensory connectivity in mice with a mEPI-NCSC graft, but not in control mice (Sieber-Blum, 2010; Hu *et al.*, 2010). Furthermore, the Semmes–Weinstein touch test showed that there was a significant improvement in touch perception. Additionally, mEPI-NCSC express angiogenic and neurotrophic factors as well as matrix metalloproteases (Hu *et al.*, 2010). Expression of the angiogenic factors, VegfA and VegfB, are likely to explain why the glial scar is vascularised in mice with a mEPI-NCSC graft. EPI-NCSC-derived metalloproteases, which can degrade various extracellular matrix molecules are likely to modulate scar formation. The ability of mEPI-NCSC to differentiate into myelinating oligodendrocytes may have additional implications, such as for the treatment of multiple sclerosis, and it further highlights the potential of mEPI-NCSC. The data gained from our studies in the mouse are provocative and therefore led us to explore the existence of equivalent human stem cells.

Human EPI-NCSC (hEPI-NCSC)

In comparison to mouse whisker follicles, human hair follicles are a simpler mini-organ, as they do not perform a sensory function. The underlying anatomy is however similar and the same phases of growth, shedding and quiescence can be observed (Paus *et al.*, 1999; Stenn & Paus, 2001). Using the same underlying principle and strategy of the migratory ability of neural crest cells, we have developed a protocol for the isolation of human equivalent cells. As in the mouse, the bulge, which in human hair has a slightly different anatomy, is dissected and explanted into adherent culture.

We have isolated hEPI-NCSC as a pure population of migratory cells from bulge explants of human hair follicles. These cells have been extensively characterised as neural crest-derived multipotent stem cells (Clewes *et al.*, 2011).

Others utilise an enzymatic digestion protocol that causes cell aggregates to form on intact cultured hair follicles (Yu *et al.*, 2006; Yu *et al.*, 2010). While the authors suggest these cells to be of neural crest origin, the cells do not show migratory activity and are more likely to be a mixture of neural crest-derived cells and epidermal stem cells, as these two cell types share the niche in the bulge. Others have also used hair follicle derived cells and shown them to be of potential benefit in severed sciatic nerve models (Amoh *et al.*, 2009). These cells are however derived from an area above the bulge and thus are unlikely to be EPI-NCSC.

The minimal invasive procedure, which we have developed for the micro-dissection of the follicular bulge is straightforward. It also relies on the migratory ability of neural crest cells and allows for elimination of other structures such as the dermal papilla, which is a known niche for SKP cells. It is an efficient method as well, as it does not rely on enzymatic treatments. Culturing of the bulge region from the anagen stage in human hair follicles alone results in a highly pure population of emigrating hEPI-NCSC (Figure 1). In contrast to mEPI-NCSC, the onset of emigration of hEPI-NCSC is somewhat later, starting between 6 and 10 days post-explantation of the bulge, after which the cells continue to migrate in the culture plate and proliferate rapidly (Clewes *et al.*, 2011; see, e.g., http://www.youtube.com/watch?v=TB-lYIPmz9I). The hEPI-NCSC have the characteristic stellate morphology of neural crest cells. Greater than 50% of cultured explants yield hEPI-NCSC (Clewes *et al.*, 2011). Due to the short time frame in culture, hEPI-NCSC are less likely to be subjected to influences that may alter their stemness or compromise chromosomal integrity.

Neural crest origin of hEPI-NCSC

As the ontological origin of a human cell type can be determined by analogy only, we determined whether hEPI-NCSC express the mouse neural

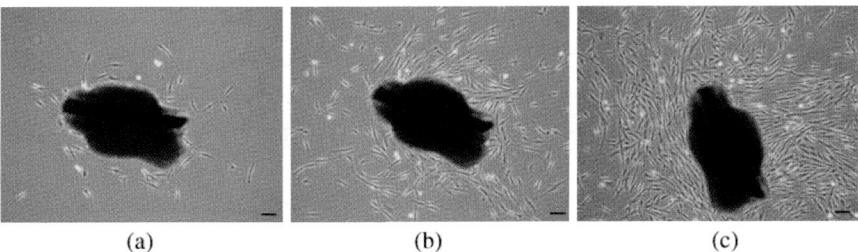

| (a) | (b) | (c) |

Figure 1. hEPI-NCSC emigrate from hair follicle bulges in primary culture. Micro dissected hair follicle bulges were explanted and incubated with growth medium. Cells were seen to emigrate from the bulge, migrate and proliferate after six, eight and 10 days respectively (a–c). Scale bars = 100 μm.

crest stem cell molecular signature. Analyses by real time PCR (qPCR) and immunocytochemistry confirmed that signature genes are indeed expressed. We were unable to analyze some of the genes, as they were RIKEN sequences in the mouse signature. We confirmed that VDAC1, ETS1, PCBP4, MYO10, H1FX, THOP1, MSX2, CRYAB, VARS2, PEG10, CALR, CRMP1, UBE4B, PYGO2, AGPAT6, and ADAM12 are expressed in all hEPI-NCSC. Additional neural crest-specific genes (SOX10, SNAI2, TWIST, MUSASHI-1, and p75NTR), global stem cell genes (TERT, NESTIN, and CD34) and the six essential pluripotency genes (C-MYC, KLF4, SOX2, LIN28, POU5F1/OCT4, and NANOG) are expressed as well at the RNA level (Clewes *et al.*, 2011). Analysis at the protein level confirmed expression of neural crest genes (SOX10 and NESTIN) and neural crest stem cell molecular signature genes ETS1, THOP1, MSX2, CRMP1, UBE4B, MYO10, ADAM12, and CRYAB (Clewes *et al.*, 2011).

hEPI-NCSC are multipotent and capable of self-renewal

Multipotency and self-renewal abilities of hEPI-NCSC were shown by *in vitro* clonal analyses. Single clone-forming cells were marked and monitored during *in vitro* development (Figure 2). The migratory behaviour of hEPI-NCSC was indicated by the changing shape of the clone. A high percentage of hEPI-NCSC, 65%, formed fast-growing colonies even under the harsh conditions of clonal culture, indicating that at least 65% of cells are multipoint stem cells (Clewes *et al.*, 2011). The remaining cells either divided once or twice only or did not proliferate and subsequently died.

Indirect immunocytochemistry was used to show that hEPI-NCSC clonal cultures can give rise to all major neural crest derivatives, including neurons (βIII-tubulin), bone/cartilage cells (collagen type II), Schwann cells [glial fibrillary acidic protein (GFAP)] and smooth muscle cells (smooth muscle actin; SMA) (Figure 3). Furthermore, serial cloning showed that cloned cells were able to self-renew and to subsequently form secondary and tertiary clones, which contained multiple cell types. This result can only be obtained when stem cells are capable of giving rise to daughter stem cells, and thus shows the self-renewal capability of hEPI-NCSC. The fact that clones contained numerous different cell types proves the multipotency of hEPI-NCSC.

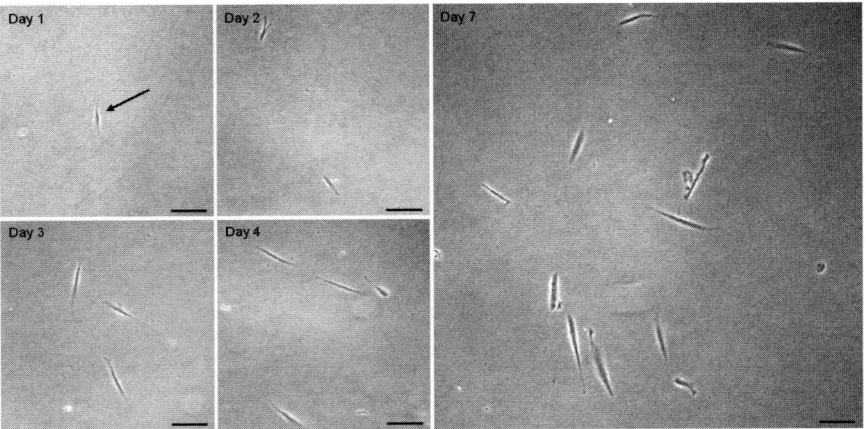

Figure 2. hEPI-NCSC form clones from a single founder cell. hEPI-NCSC were seeded at clonal density and single cells identified (day 1, arrow). Cells were circled using a diamond tipped scribe with 4 mm diameter. The same clone is shown here at two, three, four, and seven days, respectively, as labelled. Scale bars = 100 μm.

Ex vivo expansion of hEPI-NCSC

Any potential future use of hEPI-NCSC for regenerative medicine and/or cell-based therapies relies on the ability to generate sufficiently high numbers of stem cells. The hEPI-NCSC can be easily expanded *ex vivo* into millions of stem cells. On an average, 3 million cells can be obtained from one bulge explant within 28 days. Analyses at the RNA and protein levels show that all markers as listed previously continue to be expressed with few changes in expression levels (Clewes *et al.*, 2011). We also showed that *ex vivo* expanded cells remain multipotent (Clewes *et al.*, 2011).

hEPI-NCSC express the six essential pluripotency genes at RNA and protein levels

Seminal work by Yamanka and colleagues identified four factors which when exogenously expressed were capable of inducing cellular reprogramming such that cells reverted to an embryonic stem cell-like state that were termed induced pluripotent stem cells (iPS cells)(Takahashi & Yamanaka, 2006). Murine fibroblasts could be reprogrammed to iPS cells following

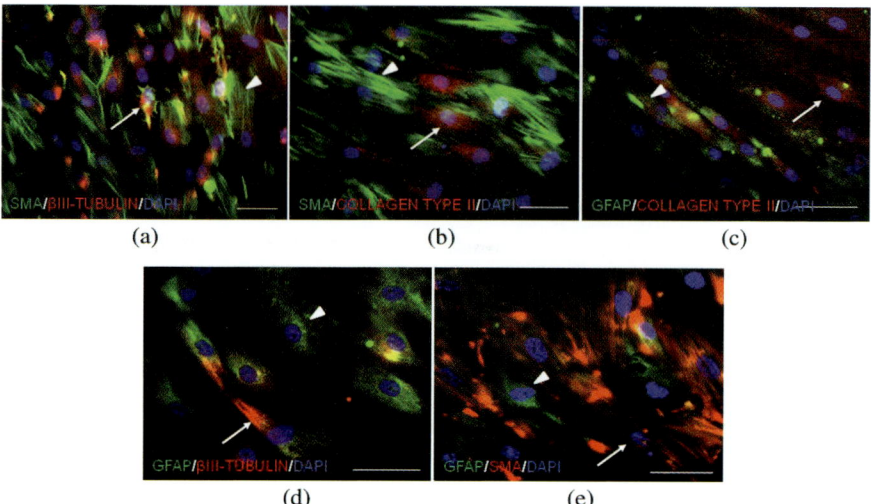

Figure 3. *In vitro* clonal analysis shows hEPI-NCSC are multipotent. hEPI-NCSC in clonal cultures show distinct cell types by immunocytochemistry triple stains. Specific antibody stains are as labelled for colour and DAPI (blue) for nucleus. (a) Smooth muscle actin (SMA) (arrow head) staining for smooth muscle cells and βIII-Tubulin (arrow) for neurons. βIII-Tubulin positive cells are also seen to have a neuronal morphology with a cell body and extending neurites. (b) Discrete cells were also positive for SMA (arrow head) and bone/cartilage cell marker collagen Type II (arrow). (c) Cells were immunoreactive for collagen type II (arrow) and the Schwann cell marker GFAP (arrow head). (d) shows subsets of cells are positive for neuronal βIII-Tubulin (arrow) and Schwann cell GFAP (arrow head). (e) GFAP (arrow head) was expressed in distinct cells from SMA (arrow). Scale bars = 50 μm.

introduction of four factors, Sox2, Klf4, c-Myc, and Oct3/4. This result introduced the concept that any somatic cell could be reprogrammed to an undifferentiated state and has since been shown to be successful in human cells. Since the initial work, research into iPS cells has snowballed with numerous types of approaches now reported. The reprogramming techniques and methodologies have also now progressed such that there is no need to use retroviruses. Other gene insertion techniques such as transposons (mobile DNA sequences containing the genes of interest) are attractive alternatives (Woltjen *et al.*, 2009) along with recombinant proteins (Zhou *et al.*, 2009) and small molecules (Yuan *et al.*, 2011). These alternative approaches negate the possible effects of viral mediated delivery and do not affect the genome of the host cells. It has also recently been reported that for neural stem cells of both murine and fetal human origin, reprogramming has been possible following the ectopic expression of POU5F1/ OCT4 alone (Kim *et al.*, 2009a; Kim *et al.*, 2009b). We have compared the expression levels of pluripotency genes to those in human embryonic stem cells (H9 line; Clewes *et al.*, 2011). C-MYC is expressed at similar levels, but transcripts for other iPS genes were lower in hEPI-NCSC when compared with H9 cells. All six pluripotency factors are expressed at the protein level as well. With some exceptions, the protein is localised in the cytoplasm rather than in the nucleus, indicating that some of the pluripotency genes exist in hEPI-NCSC in an inactive form. This is to be expected as hEPI-NCSC are an adult type of stem cell. However, the data also indicate that hEPI-NCSC are attractive candidates for reprogramming into iPS cells (Clewes *et al.*, 2011).

Directed differentiation of hEPI-NCSC into osteocytes

Osteogenic differentiation of hEPI-NCSC is of interest as many human conditions require bone transplants, ranging from osteoarthritis to difficult bone fractures and joint replacements. Cell-based therapy is easier and possibly more effective than autologous or allogeneic bone transplants. *In vitro* clonal analysis showed that hEPI-NCSC are able to express the bone/cartilage marker collagen type II. Osteogenic differentiation has been well documented for other stem cells, including mesenchymal stem cells, bone marrow mesenchymal stem cells and embryonic stem cells (Lee *et al.*, 2007; Esposito *et al.*, 2009; Agata *et al.*, 2009). Typically, osteogenic differentiation is performed through addition of factors such as dexamethasone, ascorbic acid and beta-glycerophosphate to the culture medium. Additionally, several commercially available osteogenic differentiation medium kits are increasingly available, making culturing more straightforward and controlled. Osteogenic differentiation is typically determined through detection of specific markers as described in reviews by Komori (2005; 2010). Other analyses include Alizarin Red S staining for deposition of extracellular calcium and alkaline phosphatase staining of osteoblasts.

We have shown that hEPI-NCSC can be differentiated along the osteogenic lineage (Clewes *et al.*, 2011). Surprisingly, the key transcription factors runt-related transcription factor 2 (RUNX2) and core-binding factor subunit beta (CBFB) are already expressed at the RNA, but not protein, level in hEPI-NCSC in primary explants, that is before the cells are exposed to the bone differentiation factors. Additionally, transcripts for matrix components osteopontin, osteocalcin, and collagen type I and II are also expressed. Validation at the protein level showed that differentiated osteocytes express RUNX2, albeit at lower levels, collagen type II, osteopontin, and osteocalcin (Clewes *et al.*, 2011). Finally, hEPI-NCSC-derived osteocytes are positive for Alizarin Red S, a measure for extracellular calcium deposition (Figure 4).

Directed differentiation of hEPI-NCSC into melanocytes

Neural crest cells give rise to all melanoyctes in the skin and internal organs. hEPI-NCSC can be differentiated into melanocytes with high efficiency. In Figure 5, cells are shown after DOPA reaction, which serves to augment the dark colour of melanosomes. In this preliminary protocol, greater than 65% of cells differentiated into melanocytes. In our

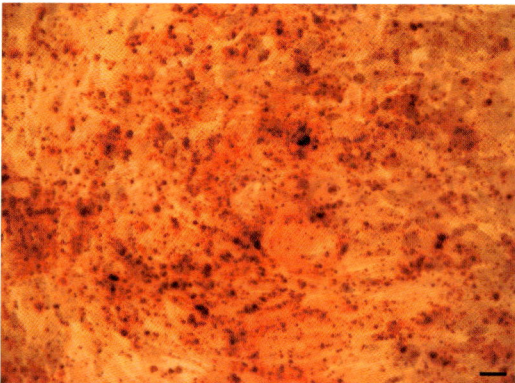

Figure 4.   Alizarin Red S staining of osteocytes derived from hEPI-NCSC. Osteogenic differentiation of hEPI-NCSC results in the extracellular deposition of calcium, seen as punctuated red areas by Alizarin Red S staining. Scale bar = 100 μm.

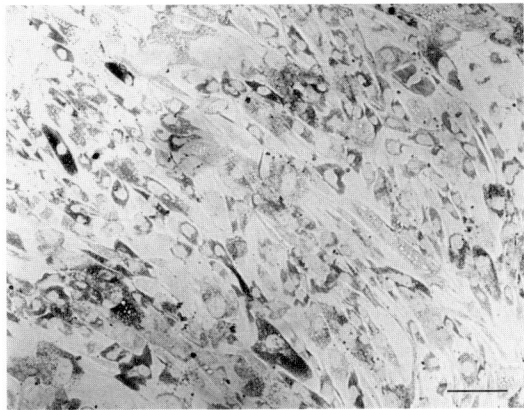

Figure 5. hEPI-NCSC express pigment granules following melanocyte differentiation. hEPI-NCSC differentiated into melanocytes results in black melanin granules within the cell cytoplasm. Scale bar = 100 µm.

current improved protocol, 99.6% of cells differentiate into melanocytes (Clewes *et al.*, unpublished). Furthermore, greater than 90% of cells were immunoreactive for the essential factor dopachrome tautomerase. This observation suggests that the cells, which were not pigmented, were melanocyte progenitor cells (Clewes *et al.*, 2011). Thus, overall, our current preliminary protocol suggests that hEPI-NCSC may become a convenient source of melanocytes, which could be used in artificial skin to treat, for instance, burn victims. One of the current limits to engineered skin is that it is often unpigmented. Thus melanocytes derived from hEPI-NCSC could be an essential component of artificial skin.

Conclusions

Our characterisation of hEPI-NCSC and validation of data suggest that hEPI-NCSC have potential for future therapeutic use. As remnants of an embryonic tissue, the neural crest, in a postnatal location, the bulge of hair follicles, hEPI-NCSC show a combination of desirable characteristics. They can be accessed in the hairy skin by minimal invasive procedures. As multipotent stem cells they have the innate ability to differentiate into a wide array of diverse cell types, which makes it straightforward to develop pertinent protocols. The hEPI-NCSC can be expanded *ex vivo* into millions of stem cells, which is an essential prerequisite for potential future therapeutic applications. So far we have developed protocols for the directed differentiation of hEPI-NCSC into osteocytes and melanoyctes (Clewes *et al.*, 2011), as well as into various types of clinically relevant neurons (Narytnyk *et al.*, unpublished). Expression of the six essential pluripotency genes at the protein level suggests that reprogramming of hEPI-NCSC might be straightforward. Last but not least, as somatic stem cells EPI-NCSC are unlikely to be tumorigenic.

References

Agata H, Watanabe N, Ishii Y, Kubo N, Ohshima S, Yamazaki M, Tojo A, Kagami H (2009). Feasibility and efficacy of bone tissue engineering using human bone marrow stromal cells cultivated in serum-free conditions. *Biochem. Biophys. Res. Commun.* 382: 353–358.

Amoh Y, Kanoh M, Niiyama S, Hamada Y, Kawahara K, Sato Y, Hoffman RM, Katsuoka K (2009). Human hair follicle pluripotent stem (hfPS) cells promote regeneration of peripheral-nerve injury: An advantageous alternative to ES and iPS cells. *J. Cell Biochem.* 107: 1016–1020.

Clewes O, Narytnyk A, Gillinder KR, Loughney AD, Murdoch AP, Sieber-Blum M (2011). Human Epidermal Neural Crest Stem Cells (hEPI-NCSC)-Characterization and Directed Differentiation into Osteocytes and Melanocytes. *Stem Cell Rev.* doi: 10.1007/s12015-011-9255-5.

Esposito MT, Di Noto R, Mirabelli P, Gorrese M, Parisi S, Montanaro D, Del Vecchio L, Pastore L (2009). Culture conditions allow selection of different mesenchymal progenitors from adult mouse bone marrow. *Tissue Eng. Part A* 15: 2525–2536.

Fernandes KJ, Toma JG, Miller FD (2008). Multipotent skin-derived precursors: Adult neural crest-related precursors with therapeutic potential. *Philos. Trans. R. Soc. Lond. B Biol. Sci.* 363: 185–198.

Hu YF, Gourab K, Wells C, Clewes O, Schmit BD, Sieber-Blum M (2010). Epidermal neural crest stem cell (EPI-NCSC)-mediated recovery of sensory function in a mouse model of spinal cord injury. *Stem Cell Rev.* 6: 186–198.

Hu YF, Zhang ZJ, Sieber-Blum M (2006). An epidermal neural crest stem cell (EPI-NCSC) molecular signature. *Stem Cells* 24: 2692–2702.

Hunt DP, Morris PN, Sterling J, Anderson JA, Joannides A, Jahoda C, Compston A, Chandran S (2008). A highly enriched niche of precursor cells with neuronal and glial potential within the hair follicle dermal papilla of adult skin. *Stem Cells* 26: 163–172.

Jiang X, Iseki S, Maxson RE, Sucov HM, Morriss-Kay GM (2002). Tissue origins and interactions in the mammalian skull vault. *Dev. Biol.* 241: 106–116.

Kim JB, Greber B, Arauzo-Bravo MJ, Meyer J, Park KI, Zaehres H, Scholer HR (2009a). Direct reprogramming of human neural stem cells by OCT4. *Nature* 461: 649–653.

Kim JB, Sebastiano V, Wu G, Arauzo-Bravo MJ, Sasse P, Gentile L, Ko K, Ruau D, Ehrich M, van den BD, Meyer J, Hubner K, Bernemann C, Ortmeier C, Zenke M, Fleischmann BK, Zaehres H, Scholer HR (2009b). Oct4-induced pluripotency in adult neural stem cells. *Cell* 136: 411–419.

Komori T (2005). Regulation of skeletal development by the Runx family of transcription factors. *J. Cell Biochem.* 95: 445–453.

Komori T (2010). Regulation of bone development and extracellular matrix protein genes by RUNX2. *Cell Tissue Res* 339: 189–195.

Le Douarin NM, Kalcheim C (1999). The Neural Crest. Cambridge and New York, Cambridge University Press.

Lee G, Kim H, Elkabetz Y, Al Shamy G, Panagiotakos G, Barberi T, Tabar V, Studer L (2007). Isolation and directed differentiation of neural crest stem cells derived from human embryonic stem cells. *Nat. Biotechnol.* 25: 1468–1475.

Paus R, Muller-Rover S, Van DV, Maurer M, Eichmuller S, Ling G, Hofmann U, Foitzik K, Mecklenburg L, Handjiski B (1999). A comprehensive guide for the recognition and classification of distinct stages of hair follicle morphogenesis. *J. Invest Dermatol.* 113: 523–532.

Sieber-Blum M (2004). Cardiac neural crest stem cells. *Anat. Rec. A Discov. Mol. Cell Evol. Biol.* 276: 34–42.

Sieber-Blum M (2010). Epidermal neural crest stem cells and their use in mouse models of spinal cord injury. *Brain Res. Bull.*

Sieber-Blum M, Grim M (2004). The adult hair follicle: Cradle for pluripotent neural crest stem cells. *Birth Defects Res. C. Embryo.* Today 72: 162–172.

Sieber-Blum M, Grim M, Hu YF, Szeder V (2004). Pluripotent neural crest stem cells in the adult hair follicle. *Dev. Dyn.* 231: 258–269.

Sieber-Blum M, Hu Y (2008). Mouse epidermal neural crest stem cell (EPI-NCSC) cultures. *J. Vis. Exp.*

Sieber-Blum M, Schnell L, Grim M, Hu YF, Schneider R, Schwab ME (2006). Characterization of epidermal neural crest stem cell (EPI-NCSC) grafts in the lesioned spinal cord. *Mol. Cell Neurosci.* 32: 67–81.

Stenn KS, Paus R (2001). Controls of hair follicle cycling. *Physiol Rev.* 81: 449–494.

Takahashi K, Yamanaka S (2006). Induction of pluripotent stem cells from mouse embryonic and adult fibroblast cultures by defined factors. *Cell* 126: 663–676.

Toma JG, Akhavan M, Fernandes KJ, Barnabe-Heider F, Sadikot A, Kaplan DR, Miller FD (2001). Isolation of multipotent adult stem cells from the dermis of mammalian skin. *Nat Cell Biol* 3: 778–784.

Toma JG, McKenzie IA, Bagli D, Miller FD (2005). Isolation and characterization of multipotent skin-derived precursors from human skin. *Stem Cells* 23: 727–737.

Woltjen K, Michael IP, Mohseni P, Desai R, Mileikovsky M, Hamalainen R, Cowling R, Wang W, Liu P, Gertsenstein M, Kaji K, Sung HK, Nagy A (2009). piggyBac transposition reprograms fibroblasts to induced pluripotent stem cells. *Nature* 458: 766–770.

Yu H, Fang D, Kumar SM, Li L, Nguyen TK, Acs G, Herlyn M, Xu X (2006). Isolation of a novel population of multipotent adult stem cells from human hair follicles. *Am. J. Pathol.* 168: 1879–1888.

Yu H, Kumar SM, Kossenkov AV, Showe L, Xu X (2010). Stem cells with neural crest characteristics derived from the bulge region of cultured human hair follicles. *J. Invest Dermatol.* 130: 1227–1236.

Yuan X, Li W, Ding S (2011). Small molecules in cellular reprogramming and differentiation. *Prog. Drug Res.* 67: 253–266.

Zhou H, Wu S, Joo JY, Zhu S, Han DW, Lin T, Trauger S, Bien G, Yao S, Zhu Y, Siuzdak G, Scholer HR, Duan L, Ding S (2009). Generation of induced pluripotent stem cells using recombinant proteins. *Cell Stem Cell* 4: 381–384.

6

NEURAL CREST STEM CELLS FROM THE HEAD REGION

Christian Kaltschmidt and Barbara Kaltschmidt

Cell Biology, University of Bielefeld, Universitätsstr. 25, D-33501 Bielefeld, Germany
Molecular Neurobiology, University of Bielefeld, Universitätsstr. 25, D-33501 Bielefeld, Germany

Neural crest cells can be found in various adult tissues. They have a great developmental potential that can be surpassed only by the embryonic stem cells (Shakhova & Sommer, 2010).

In this chapter, we will present our results on human and rodent Neural Crest Stem Cells in light of the existing literature. Special emphasis will be laid on the isolation and culture, the endogenous niche and potential clinical applications.

Location of neural crest stem cells in the human head

The head is anatomically the most sophisticated part of our body and its evolution was fundamental to the origin of vertebrates. The head structure was introduced during the development of *Chordata* which includes vertebrates, tunicates, and cephalochordates. The development of head structure depends on the contribution of neural crest cells. The neural crest cells originate at the most dorsal region of the neural tube. The fate of the neural crest cells depends mainly on where the cells migrate and settle.

Table 1.   Derivatives of neural crest.

Head	Trunk
Neural derivates:	*Neural derivates:*
Sensory neurons, autonomic and enteric	Sensory neurons, autonomic and enteric
Glia (e.g. Schwann cells)	Glia (e.g. Schwann cells)
Melanocytes	*Melanocytes*
Endocrine cells	*Endocrine cells*
Mesenchymal derivates:	
Chondrocytes, Osteocytes	
Myofibroblasts, Smooth muscle cells, Adipocytes,	
Connective tissue cells	
Meninges (Forebrain)	
(Le Douarin *et al.*, 2008; *Nagoshi et al.*, 2008)	

Defects within the neural crest compartments result in severe malformation of the head from mice and humans, e.g. Twist (failure of neural tube closure, Saethre-Cotzen Syndrome), Tcof1 (neural crest apoptosis, Treacher Collins-Franceschetti Syndrome), Pax9 (cleft secondary palate, absent teeth, Oligodontia), Pax3 (neural tube defect, deficiency of Schwann Cells and dorsal root ganglia, Waardenburg Syndrome type 1), TGFβ2 (cleft palate, defects in neural crest skeletogenesis). For review see Wilkie & Morriss-Kay, 2001.

Stem cells are self-renewing with the capacity to generate multiple differentiated derivatives (Morrison *et al.*, 1997). In developing embryos, stem cells construct tissues and organs *de novo*, while in adults they maintain ongoing cellular turnover and provide regenerative capacity after lesion in certain tissues. Neural crest cells behave *in vitro* as multipotent self-renewing (Trentin *et al.*, 2004) stem cells/progenitors (neural crest-derived stem cells (NCSCs)) that can differentiate in multiple lineages (Sieber-Blum & Cohen, 1980; Stemple & Anderson, 1992). Lineage tracing with either quail-chicken chimera (Le Douarin & Teillet, 1974; Bronner-Fraser *et al.*, 1980) with genetically tagged neural crest cells by Wnt1,

Sox10, tPA, protein P0, promotor driven Cre Recombinase showed multi-potency *in vivo* also (Dupin *et al.*, 2009).

Recently, NCSCs of the head region were described in a number of adult organs and tissues like skin, cornea, hair follicles, periodontal ligament, hard and soft palate, pulpa of teeth, turbinate and ensheathing cells of the olfactory epithelium (Figure1) (Shakhova & Sommer, 2010; Barraud *et al.*, 2010).

Genetic lineage tracing with Wnt1[CRE]; R26R[YFP]showed that mouse neural crest cells formed olfactory ensheathing cells: The Schwann cells

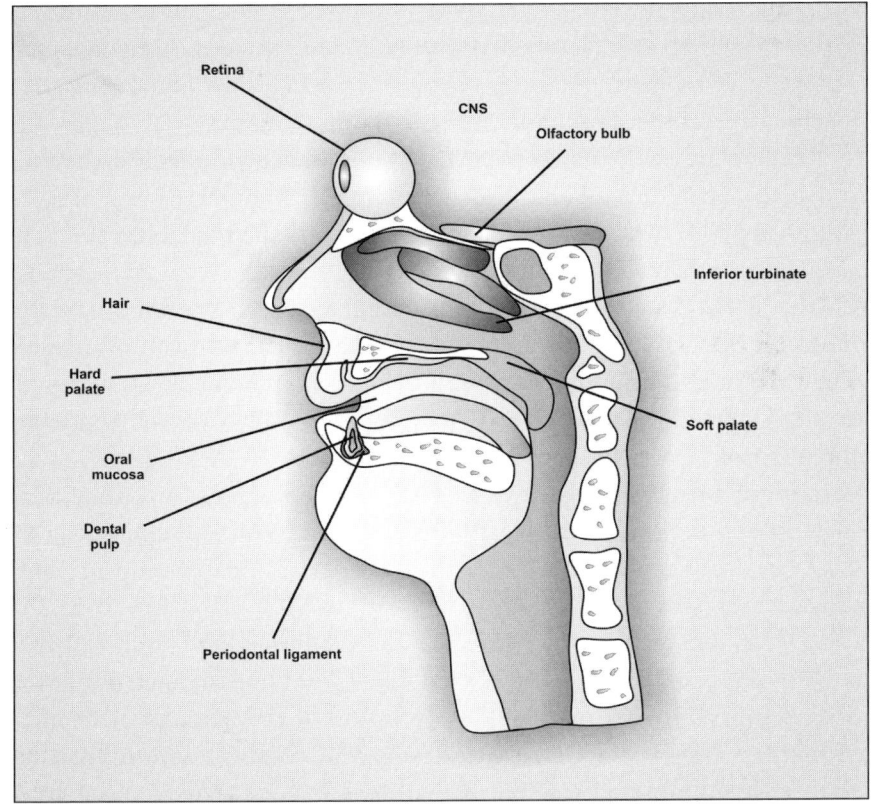

Figure 1. Scheme of the location of neural crest stem cells (NCSCs) within the human head region.

Note: Novel location of NCSCs within the inferior turbinate of the nose. For characterisation of these cells, refer to section on Nasal turbinate.

that ensheathed the olfactory nerve (Barraud *et al.*, 2010). In contrast olfactory placodes did not form olfactory ensheathing cells. Moreover olfactory ensheathing cells were positive for the established neural crest markers p75^{NTR}, P0 and Sox10.

Human neural stem cells from the periodontal ligament (pdNSCs)

Recently we were able to isolate neural stem cells from adult human tissue, which were isolated by minimally invasive periodontal surgery (access flap) (Widera *et al.*, 2007). These cells could be cultivated as spheroid cultures in the presence of growth factors FGF and EGF. Such three-dimensional neurospheres can be kept in culture for several passages without losing their capacities for proliferation, migration, and differentiation.

To characterise the newly propagated population of pdNSCs, the protein expression of several neural stem cell-specific biomarkers was assayed using antibodies; these included stemness markers such as Nestin and Sox-2. Most of the pdNSCs were anti-Nestin immunoreactive, a characteristic of NSCs. The cells also expressed the neural stem cell-specific transcription factor Sox-2. No expression was detected for the population markers L1 or LeX. In addition, no expression of markers specific for hematopoietic stem cells CD34 and CD117 was detected. We observed moderate expression of PSA-neural cell adhesion molecule (NCAM), a marker for migrating neuronal precursor cells. There was no expression of the neuronal differentiation marker β-III-tubulin and very low expression of GFAP (glial cell marker). We also studied the RNA expression of some lineage markers using RT-PCR.

CD45, a marker for blood cells, was detected in human adult CD 133^{+} hematopoietic stem cells and in RNA derived from human liver, but not in pdNSCs from different origins. No expression of the hematopoietic stem cell markers CD34 and CD90 was detected in pdNSCs. The marker for primitive lymphoid and myeloid hematopoietic bone marrow progenitor cells and mesenchymal stem cells, CD117 (stem cell factor receptor/c-kit), was only weakly expressed in pdNSCs. In contrast, the NSC markers Nestin and Sox-2 were strongly expressed. Surprisingly, the pre-oligodendrocytic marker 2′:3′-cyclic nucleotide 3′-phosphodiesterase (CNPase) was expressed in all the samples from human organs (adult brain, fetal brain, and adult liver) and in CD133 stem cells and pdNSCs. Unlike CD133$^+$ hematopoietic stem cells, pdNSCs express Emx2, an early marker for developing neocortex. There was no detectable expression of the differentiation markers GFAP for glial cells or neurofilament-3 (NF-H/NF-200kD) for neurons. Interestingly, in contrast to brain, liver, or even CD133 HSCs, pdNSCs do not express geminin (GMNN), a well-described negative regulator of replication, indicating that the cells are highly proliferative. In addition, the cells were negative for Notch1 and CD133. The marker for embryonic stem cell pluripotency, Oct-4, was not expressed in pdNSCs. Additionally, FACS analysis demonstrated expression of A2B5 (neuron cell-surface antigen), a marker for neuronal lineage, where pdNSCs were negative for PSA, LNGFR, CD29, CD73, and CD117. No expression of hematopoietic stem cell markers CD34, CD45, and CD133 was detected. Periodontal ligament is built from neural crest cells during development. In our cultures we could not detect the expression of typical neural crest markers such as p75^{NTR} (LNGFR) in serum-free neurosphere medium with EGF and FGF. In contrast, Coura and colleagues reported the expression of

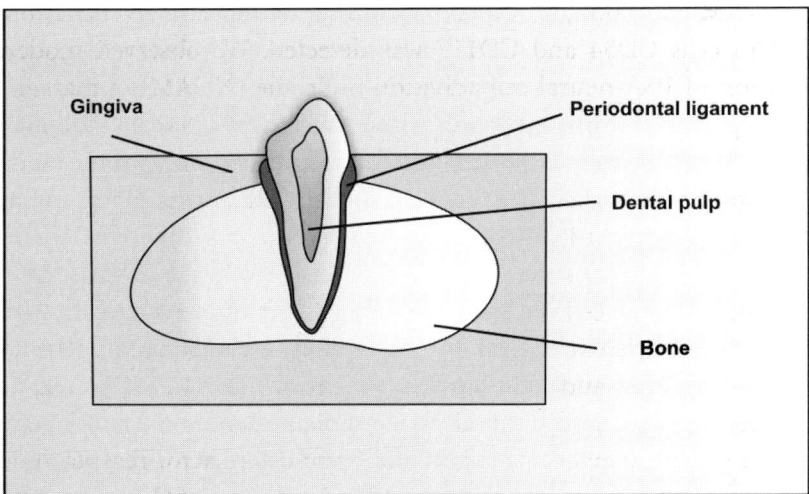

Figure 2. Schematic drawing of a human tooth, embedded in jaw-bone. NCSCs could be isolated from gingiva (oral mucosa), periodontal ligament, and dental pulp.

neural crest cell markers p75^{NTR} and HNK-1 in adherent cultures with neural crest inductive media containing, FCS, chicken embryo extract, transferrin, hydrocortisone, glucagon, insulin, triiodothyronine, epidermal growth factor and fibroblast growth factor 2 (Coura *et al.*, 2008).

Thus culture conditions seem to be an essential factor when comparing marker gene expression. In cell culture, under serum-free conditions, human pdNSCs form spheres, which are capable of producing extracellular matrix (Arnold *et al.*, 2010). Furthermore these cells could be differentiated into osteogenic lineages. In addition pdNSCs could be differentiated in glial and neuronal lineages (up to 30% of cultured cells). To enhance the frequency of neuronal cells, a differentiation protocol for somatic NSCs was developed (Widera *et al.*, 2009). Cultures were kept as aggregates in the presence of growth factors (FGF-2 and EGF) and 5 µM retinoic acid (RA) for 4 days. After dissociation, the cells were plated and analyzed. Twenty-four hours after plating, the frequency of neuronal marker β-III-tubulin-positive cells was greatly enhanced to 98.7% in comparison to 34.5% marker positive cells without RA treatment. After RA treatment these neurons expressed markers as Neurofilament M, Neurofilament H, Map-2, GAD67, Synaptophysin, and Neurofilament L.

Interestingly pdNSCs, transfected with GFP, survive and integrate when transplanted into organotypic hippocampal brain slices (Widera *et al.*, 2007). After 13 days of culture, GFP-positive cells remained detectable within the slice. Immunocytochemical analysis of transplanted, GFP-expressing cells confirmed that pdNSCs are able to spontaneously differentiate into proper lineages as demonstrated by expression of neuronal markers (NF-L, β-III-tubulin, Synaptophysin) and glial markers (GFAP) as well.

Transplantation of cultivated human pdNSC pre-differentiated with dexamethasone to an osteogenic lineage, into nude rats resulted in the formation of solid tumors (ca. 6 g weight) after six weeks (Kaltschmidt *et al.*, unpublished data), arguing against a therapeutic use of stem cells derived from the periodontal ligament.

Taken together, neural stem cells from the periodontal ligament might have several disadvantages as a potential stem cell source for therapy:

1. Long time expansion in tissue culture is necessary to obtain enough cells for a replacement therapy due to the small samples extracted by

minimally invasive periodontal surgery. Long time culture potentially selects for aneuploid cells with growth advantages and a high risk of tumour formation as we have seen in our experiments.

2. Absence of pluripotency markers as Oct-4, Klf-4, c-myc.

Therefore we screened for alternative sources of tissues which are easily accessible in larger quantities.

Adult palatum as a novel source of neural crest-derived stem cells from rats and humans (pNC-SCs)

Recently, we have identified cells positive for Nestin, a neural stem cell/neural crest specific marker protein, adjacent to Meissner Corpuscles (MC) and Merkel cell-neurite complexes (Me) within palatal ridges (*palatal rugae/rugae palatinae*) (Widera *et al.*, 2009).

This was the first report on mammalian Nestin-positive cell population associated to MC. Double-staining of sections against Neurofilament M (NF-M) and Nestin demonstrated that Nestin-positive cells within MC are mainly localised at the summit of the ruga, whereas the Nestin-filaments of Me are located in the deeper aspect of the rugal wall. The nerve fibres, visualised via NF-M staining, are partially co-localised with the Nestin-filaments in the central area of MC. At the summit of MC a high amount of Nestin, but only few Neurofilaments could be detected (see Figure 3).

The secondary palate is a highly regenerative and richly innervated craniofacial tissue, which develops under direct contribution of neural crest cells. It is well described that wounds within oral mucosa heal rapidly and without scar formation (Kahnberg & Thilander, 1982). This capability to rapidly regenerate may be explained either by the presence of growth factors, e.g. basic fibroblast growth factor in the saliva or by the potential presence of at least one stem cell type within the tissue.

In vitro, proliferating cells often escape from growth control as a result of genetic abnormalities that manifest in alterations including ploidy and mutations. In contrast to neurosphere cultures derived from periodontal ligament, pNCSCs harbour normal DNA content and chromosome number. After 21 passages, pNCSCs still contain 42 chromosomes typical for a Wistar rat, suggesting that pNCSCs are genetically stable.

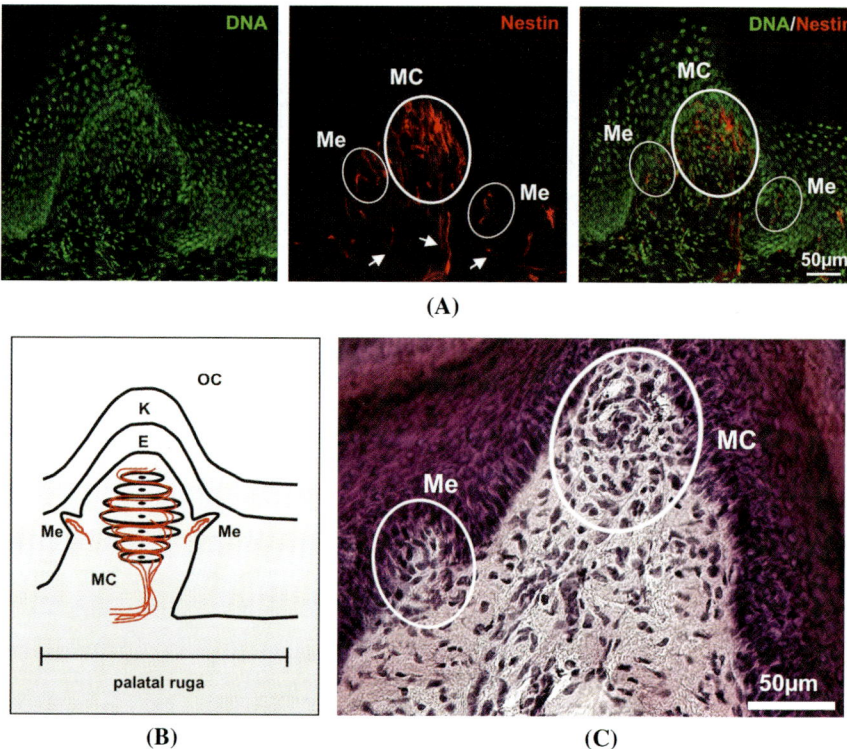

Figure 3. pNCSCs positive for the stem cell marker protein Nestin, were located adjacent to Meissner corpuscles (MC) and Merkel cell-neurite complexes (Me) within rat palate (Widera *et al.*, 2009), used from the open access version of this article). (A) Sagittal section along a palatal ruga, stained with antibodies against Nestin (red), nuclei were stained with Sytox green. *Note*: Nestin-positive cells are localised mainly within MC and Me, some cells are detected in the neighbourhood of potential nerve fibres (arrows). Bar 50 µm. (B) Schematic drawing (MC: Meissner Corpuscle, Me: Merkel cells, OC: oral cavitiy, E: epidermis with keratinised *stratum corneum* (K)). (C) HE staining of sagittal sections along palatal ruga.

In the newly propagated population of pNCSCs, the protein expression of neural stem cell specific biomarkers Nestin and Sox2 was assayed using antibodies (Figure 4A).

30.95% + 14.7 of the pNCSCs were anti-Nestin immunoreactive, a characteristic of neural/neural crest stem cells. 84.8% of the cells also expressed the neural/neural crest stem cell specific transcription factor

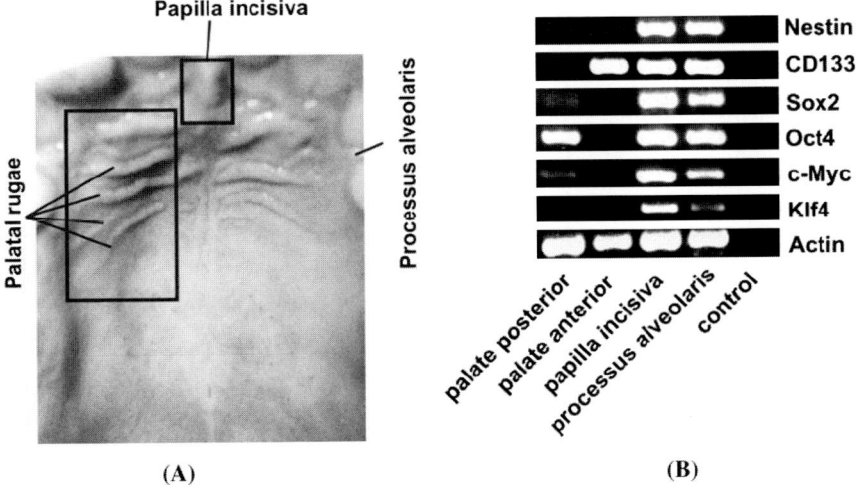

(A) (B)

Figure 4. Human palatal tissue containing pNCSCs. (A) Overview photograph of human palatal structures. (B) After tissue isolation and reverse transcription of the RNA the expression pattern of various palatal regions was investigated by PCR. Beside anterior and posterior samples from palate, samples from the papilla incisiva and the distal processus alveolaris were investigated. The highest expression of the human stem cell markers Nestin and CD133 was detected within the papilla incisiva and processus alveolaris. In additon, the tissue from this region showed high expression of all of the so called "magic four" reprogramming factors: Sox2, c-Myc, Klf4, and Oct4. cDNA was normalised using the β-actin housekeeping gene (Taken from Widera *et al.*, 2009, used from the open access version of this article).

Sox2. There was no expression of the neuronal differentiation marker β-III-tubulin. In contrast, few cells showed an expression of GFAP (glial cell marker). Additionally we studied the RNA expression of several markers using RT-PCR (Figure 4B). The stem cell markers Nestin and Sox-2 were strongly expressed in all tested samples indicating a presence of putative stem cells in all examined tissues and organs. RT-PCR verified the expression of GFAP.

The pNCSCs demonstrated high expression of neural crest markers as p75^{NTR}, ABCG2, Slug, Twist, and Sox9. Interestingly, the RNA for Slug was absent in all CNS-related control samples (cortex and cerebellum). In addition we could show a weak expression of Klf4 and c-Myc in the adult cerebellum but not in the adult cortex. As shown by *in situ* hybridisations for Klf4 this gene is transcribed in mouse adult cerebellar granule cells and not in cortex, whereas c-Myc transcription is restricted to Purkinje Cells. In summary, our PCR results match with the data gained by *in situ* hybridisations against Klf4 and c-Myc as shown in the Allen Brain Atlas (http://www.brain-map.org) where pNCSCs were positive for Notch1. Notch-signalling has been described to be crucial for cranial neural crest development and homeostasis (Cornell & Eisen, 2005). Since Sox2, one of the so-called "magic four" (Takahashi & Yamanaka, 2006) was highly expressed in pNCSCs, we tested the potential expression of the other gene products Klf4, Oct4, and c-Myc as well. Here we showed that pNCSCs were not only positive for Sox2 but for c-Myc and Klf4 as well. Demonstrably, both palatal tissue and cultivated pNCSCs expressed Oct4 (Figure 4D). Oct4 has been recently shown to be solely sufficient for reprogramming of adult neural stem cells (Kim *et al.*, 2009; Widera *et al.*, 2006).

In addition pNCSCs expressed heterogeneously distributed NF-κB. Active NF-κB is one of the crucial transcription factors for the proliferation of neural stem cells (Widera *et al.*, 2006; Wu *et al.*, 2000). Beside cells with cytoplasmic NF-κB (61.93%), several cells with nuclear NF-κB (38.05%) could be detected.

In order to investigate the differentiation potential of pNCSCs, the cultures were initially cultivated as aggregates in the presence of growth factors (FGF-2 and EGF) and 5 μM retinoic acid (RA) for four days. After at least 96 h, the spheres were dissociated and single cells were plated on poly-D lysine/laminin-coated cover slips in the absence of growth factors. After a day of plating, a neuron-like morphology (bipolar cell bodies with elaborated processes) was already detectable. Four days after plating on poly-D lysine/laminin-coated cover slips, the cells were fixed and processed for immunocytochemistry. Here we detected high frequency of cells positive for neuronal markers. 26.1% + 11.9 of the cells expressed the early neuronal marker tubulin (TuJ). Furthermore, 21% + 17.9 of differentiated pNCSCs were immuno-positive for the neuron specific intermediate filament Neurofilament-M (NF-M). 22.6% + 10.93 of the cells showed high expression of MAP2. Taken together, the differentiation capacity of pNCSCs provides a further evidence for the stemness of this cell population.

In order to identify the potential stem cell pools within human palate we investigated anterior and posterior samples from hard palate.

Additionally, samples from the papilla incisive, and the distal processus alveolaris maxillae were investigated (Figure 4A). Using RT-PCR we demonstrated the highest expression of the human stem cell markers CD133 and Nestin within the papilla incisiva and the distal processus alveolaris. In addition this population expressed high levels of Sox2, Klf4, Oct3/4, and c-Myc (Figure 4B).

While pNCSCs are a cell population of potential clinical interest, it was, in our hands, difficult to cultivate human pNCSCs from clinical material without yeast contaminations. About 75% of the cultures were contaminated even in the presence of fungicides. In addition, cultures without contamination showed a slow doubling time and could not generate secondary neurospheres, which is in sharp contrast to our observations with rat pNCSCs. Other groups reported the isolation of a multipotent neural crest-derived progenitor cell population from buccal oral mucosa lamina propria (Davies *et al.*, 2010). Within this tissue the following pluripotency markers were detected by PCR: Oct-4, Nanog, Sox2, Klf-4, and hTERT. Isolated cells cultivated in a medium containing 10% FCS formed colonies with cells depicting a fibroblast like morphology. Cultivated cells expressed the neural crest markers Snail, Slug, Sox10, and Twist. Oral mucosa lamina propria progenitor cells could be differentiated into different lineages: neurons, astrocytes, Schwann cells, mesenchymal cells as osteoblasts and chondrocytes.

Another study described the isolation of stem cells from oral mucosa lamina propria-derived from gingival and alveolar mucosa (Marynka-Kalmani *et al.*, 2010). Cells were cultivated in medium with 10% FCS by explantation of mucosa. Population doubling time was 49 hours. More than 65% expressed the pluripotency related hESC markers: SSEA4, Oct-4, and Sox-2, 40% expressed Nanog. In addition, Tra 2-54 and Tra49 were expressed similar to some ESC lineages. Alkaline phosphatase expression was two-fold higher than in hESCs. Quantitative RT-PCR revealed that Oct-4, Sox-2, and Nanog were expressed several hundred-fold higher than baseline but an order of magnitude lower than in ESCs. The p75 [NTR] expression suggests a neural crest origin of these cells. These cells could be differentiated in neural, mesodermal, and endodermal lineages. Within its endogenous niche — the lamina propria of the mucosa — chord like structures positive for p75 [NTR], Oct-4, and Sox-2 could be detected *in vivo*.

Whereas neural crest-derived stem cells could be isolated from several regions of the oral mucosa such as buccal, gingival, and palatal locations, it remained to be clarified which cell types might represent the putative stem cell/progenitor in their endogenous niche. In quails, lineage analyses using single clones of neural crest cells suggested that this cell type might generate progeny such as glia (Schwann cells), neurons and melanocytes, or mesenchymal derivates such chondrocytes, osteocytes, or fibroblasts (Le Douarin *et al.*, 2008).

To analyse this problem in the palatum of rats, we performed immuno-electron microscopy with anti-Nestin antibodies. Nestin is a marker for neural stem cells (Lendahl *et al.*, 1990) and seems to be necessary for proper self-renewal (Park *et al.*, 2010).

Anti-Nestin immunoreactivity was localised at the ultrastructural level (by silver-intensified quantum dots) exclusively within the myelinating regions of Schwann cells, i.e. within the myelin layers (Fig. 5 B, arrows) ensheathing an axon. Nestin expressions was not observed within the cytoplasmic part of Schwann cells which ensheathes unmyelinated axons (arrowheads). Blow-up visualised anti-Nestin labeling in myelinating membranes of Schwann cells ensheathing a single axon. The axon itself remained unlabelled. Schwann cell soma was not immuno-reactive to anti-Nestin, as well as the Schwann cell nucleus. Other cells within the palate such as keratinocytes, fibroblasts, and endothelial cells were negative for Nestin. These results clearly show that Schwann cells represent the putative stem cell in the palatal endogenous niche.

Extremly pure Schwann cell cultures can be cultivated from adult sciatic nerves (rat) as neurospheres. Interestingly, these neurospheres are multipotent, can differentiate in all three germ layers and show signs of reprogramming such as the expression of pluripotency factors (c-Myc, Oct-4, Klf4, Sox2) (Widera & Kaltschmidt, unpublished). In contrast, isolated sciatic nerve showed no expression of pluripotency factors at all.

While the Schwann cells myelinate axons in the PNS, the oligodendrocytes myelinate axons within the CNS. Previously it was shown, that the CNS oligodendrocytic precursors could be reprogrammed by culture to multipotent CNS stem cells (Kondo & Raff, 2000). Taken together, these data suggest that myelinating cells such as Schwann cells and Oligodendrocytes have, in addition to their role in myelination, an important function in tissue regeneration.

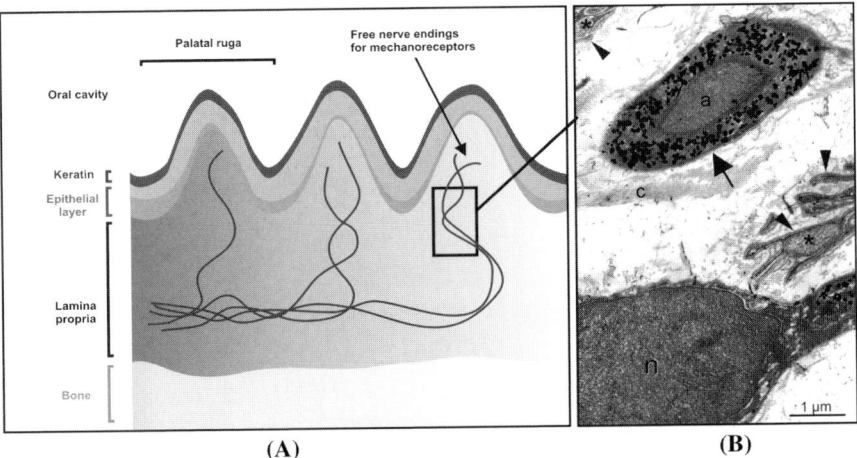

(A)   (B)

Figure 5. In cross sectioned nerves of palatal rugae, Nestin is exclusively localised at the ultrastructural level (by silver-intensified quantum dots) within the myelinating regions of Schwann cells, i.e. within the myelin layers (arrows) ensheathing an axon. Nestin was never found within the cytoplasmic part of Schwann cells which ensheathes unmyelinated axons (asterix). Magnification ×18.000. a: myelinated axon; c: collagen; f: fibroblast; n: nucleus of Schwann cell; asterix: unmyelinated axon; arrowheads: non-myelinated region of Schwann cell; arrows: myelinated part of Schwann cell.

Within the quail system, Schwann cells could be de-differentiated to lineage restricted precursors which generate glial and melanocytic progeny by treatment with the cytokine endothelin 3 (Dupin *et al.*, 2000; Dupin *et al.*, 2003; Raff, 2003).

Nasal turbinate

Olfactory ensheathing cells are a glial cell type associated with the axons of olfactory sensory neurons (Figure 6). Despite its proven therapeutic potential (see below) the isolation of olfactory ensheathing cells involves a sophisticated surgery technique since these cells are part of the central nervous system. Recently we succeeded in the isolation of potential human neural crest-like stem cells from the lower part of the human nose (Figure 1, inferior turbinate), which were named inferior turbinate stem cell (ITSC). ITSCs could be detected in the tissue within the lamina propria of respiratory mucosa using an anti-Nestin antibody (Figure 6B). Epithelial tissue is

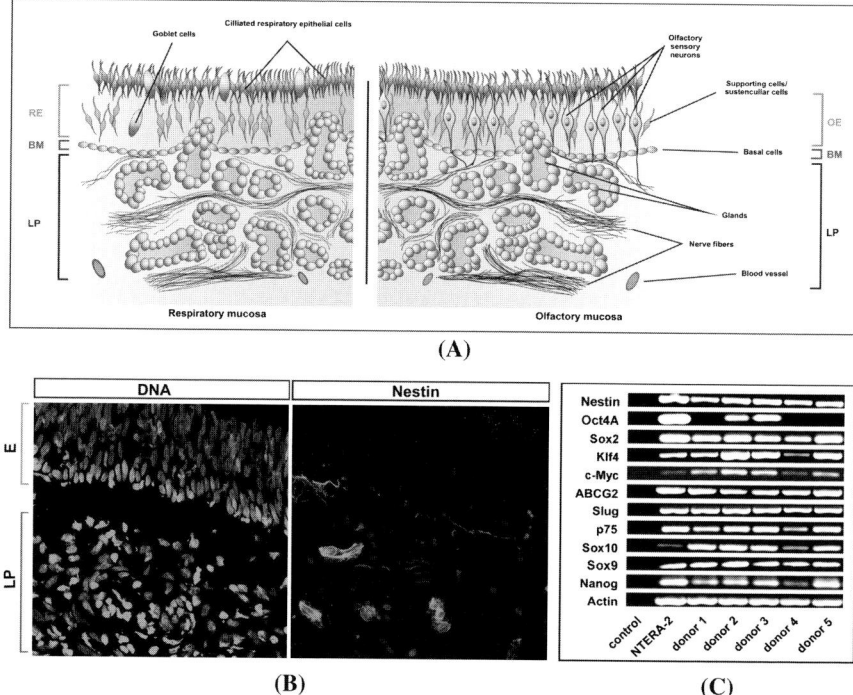

Figure 6. Localisation of ITSCs within the human nasal mucosa and their characterisation. (A) Schematic drawing of human olfactory and respiratory mucosa. (B) Nestin staining of human inferior turbinate revealed that ITSCs are located within lamina propria (LP) of respiratory epithelium and are distinct from basal cells within the epithelial layer (E), which are only weakly positive for Nestin. (C) RT-PCR-analysis of the adult human inferior turbinate tissue. Robust expression of neural crest markers Sox 10, Sox 9, p75 and Slug was demonstrated in five of five analyzed donors. In addition, the turbinate tissue of all donors was positive for Sox2, Klf4, cMyc, Nanog. Moreover, three of five tissue probes were positive for Oct4.

only sparsely labelled, indicating that ITSCs are non-epithelial. Analysis of neural crest markers Sox 10, Sox 9, p75[NTR] and Slug showed a robust expression in five of five analyzed donors. ITSCs seem to have a great developmental potential, since transcription factors usable for forced reprogramming were expressed already (Oct4, Sox2, Klf4, cMyc, Nanog) in a freshly isolated tissue. ITSCs show robust self-renewal, could be isolated free of contaminations and could be differentiated into multiple lineages, representing all three germ layers. Taken together, ITSCs seem to be an ideal source for cranial neural crest-like stem cells for cell replacement therapy.

Pre-clinical and clinical research

Skin stem cells (holoclones), Barrondon & Green, 1987) could be used for human therapy of life threatening burns or cornea regeneration (Pellegrini *et al.*, 2009) suggesting that adult stem cell therapy is feasible.

As summarised above, neural stem cells can be cultivated from various sources of adult humans making this highly plastic cell type available for therapeutic trials. Here we review pre-clinical and potential clinical applications of neural crest-derived stem cells from the head region. The developmental potential of cranial neural crest cells suggests that they might be useful for regeneration of neuro-ectodermal tissue such as PNS/CNS neurons and Schwann cells. Furthermore, reconstruction of bone and cartilage might be feasible.

Spinal cord

Interestingly, animal experiments use EPI-NCSCs for treating spinal cord lesions. EPI-NCSCs can be easily isolated from hairy skin and express the pluripotency markers c-Myc, Klf-4, Sox2, Lin28 (Sieber-Blum *et al.*, 2004). The Semmes–Weinstein touch test measures sensory perception which is hampered in hind limbs after spinal cord lesion. Surprisingly, transplantation of EPI-NCSCs reverted the sensory defect to control values (Sieber-Blum, 2010; Hu *et al.*, 2010). In contrast to undifferentiated embryonic stem cells they do not form tumours in the spinal cord (Sieber-Blum *et al.*, 2006). EPI-NCSCs could generate oligodendrocytes and GABA-ergic neuroblasts.

Olfactory ensheathing cells surround olfactory axons into the CNS. Recently it was shown that the olfactory ensheathing cells cultured from adult olfactory bulb, expressing p75^{NTR}, could improve a corticospinal tract lesion (Li *et al.*, 1997).

Human spinal cord

Olfactory ensheathing cells — another neural crest-derived cell type — were already used for cell transplantation in human spinal cord injury. The source for autologous olfactory ensheathing cells is the olfactory mucosa, which can be isolated by neurosurgery from the superior region of the nasal cavity close to the cribriform plate with the help of a nasal endo-scope. After transplantation of olfactory ensheathing cells, one out of three patients showed a moderate recovery in light touch and pin prick sensitiv-ity (Mackay-Sim *et al.*, 2008). There are three more studies reporting improvement in motor and sensory scores, unfortunately all did not use controls (King-Robson, 2011). Some caution might be necessary since cell transplantation can also have adverse neurological effects e.g. in patients with incomplete spinal cord injury, in addition sample size might be too small for meaningful statistical analysis. While benefits of stem cell trans-plantation after spinal cord lesions have been demonstrated, complete functional recovery could not be shown so far.

Bone repair

Another cell-type which is similar to neural crest stem cells are skin-derived precursors (SKPs) from hair follicles (Jinno *et al.*, 2010).

Surprisingly, SKPs derived from the trunk region are of a somite ori-gin, but SKPs from the head and trunk are functionally very similar. They are both multipotent stem cells and give rise to peripheral neural and some mesodermal cell types, such as adipocytes and share an overlapping gene expression profile. Rodent and human SKPs can generate mesenchy-mally derived cell types, such as osteocytes and chondrocytes. Clonal analysis showed that SKPs generated skeletogenic cell types, which were multipotent with regard to the osteogenic and chondrogenic lineages. Furthermore, within a tibial bone fracture model, transplanted SKPs could regenerate bone fractures (Lavoie *et al.*, 2009). Recently it was shown that SKPs are adult dermal stem cells, which can regenerate even hair, when transplanted to a host animal (Biernaskie *et al.*, 2009).

Craniofacial reconstruction

In Europe about 1.5 million patients need craniofacial reconstruction each year. Nearly 20% of these patients suffer from loss of function due to unsuccessful treatments. The closure of bone defects is commonly treated by transfer of tissue (e.g. flap), which might not fully restore tissue integ-rity. One of the main applications of craniofacial reconstruction is the repair of aesthetic and functional defects in cases of congenital malformations, tumour resections and traumatic injuries.

So far several studies have been published on the use of stem cells in craniofacial reconstructions. A stem cell formulation called injectable bone (cultured mesenchymal stem cells pre-differentiated into osteoblast lineage, mixed with polymerized fibrin) was used for articulate reconstruction with longterm success (Ueda *et al.*, 2008). In another study, human mandible bone defects were repaired by grafting collagen sponges seeded with dental pulp-derived stem cells (d'Aquino *et al.*, 2009). In some clinical situations, wisdom tooth extraction generates a critical bone defect, which does not allow bone repair with classical surgery techniques. Therefore, extracted maxillary molars were used to isolate the pulps. These were dissociated and cultured with 20% FBS for 21 days. No selection procedure for stem cells was used. Transplantation was done with collagen sponges injected with cultured cells. Formation of new bone and efficient healing could be verified with radiology and with analysis of extracted bone tissue.

Parkinson's disease

Parkinson's disease is a movement disorder due to degeneration of dopaminergic neurons within the substantia nigra. Thus a cellular replacement therapy might be feasible and desirable. Cultured human OECs from healthy controls and Parkinson's disease patients were used in the 6-hydroxy-dopamine animal model. The beneficial effects of transplanted OECs were seen in both the healthy and PD patients. After stem cell therapy, amphetamine induced rotational bias was eliminated completely. No tumour formation was detected. It is noteworthy that dopaminergic differentiation was only seen after culturing OECs as neurospheres (Murrell *et al.*, 2008).

Taken together, human neural crest stem cells from the head region have a wide developmental potential even in situations where a functional recovery might ask for cells from a different germ layer such as CNS repair.

Acknowledgement

We thank Darius Widera for the illustrations and critical reading.

References

Arnold WH, Becher S, Dannan A, Widera D, Dittmar T, Jacob M, Mannherz HG, Kaltschmidt B, Kaltschmidt C, Grimm WD (2010). Morphological characterization of periodontium-derived human stem cells. *Ann. Anat.* 192(4): 215–219.

Barrandon Y, Green H (1987). Three clonal types of keratinocyte with different capacities for multiplication. *Proc. Natl. Acad. Sci.* USA 84(8): 2302–2306.

Barraud P, Seferiadis AA, Tyson LD, Zwart MF, Szabo-Rogers HL, Ruhrberg C, Liu KJ, Baker CV (2010). Neural crest origin of olfactory ensheathing glia. *Proc. Natl. Acad. Sci.* USA, 107(49): 21040–21045.

Biernaskie J, Paris M, Morozova O, Fagan BM, Marra M, Pevny L, Miller FD (2009). SKPs derive from hair follicle precursors and exhibit properties of adult dermal stem cells. *Cell Stem Cell* 5(6): 610–623.

Bronner-Fraser M, Sieber-Blum M, Cohen AM (1980). Clonal analysis of the avian neural crest: Migration and maturation of mixed neural crest clones injected into host chicken embryos. *J. Comp. Neurol.* 193(2): 423–434.

Cornell RA, Eisen JS (2005). Notch in the pathway: the roles of Notch signalling in neural crest development. Semin. *Cell Dev. Biol.* 16(6): 663–672.

Coura GS, Garcez RC, de Aguiar CB, Alvarez-Silva M, Magini RS, Trentin AG (2008). Human periodontal ligament: A niche of neural crest stem cells. *J. Periodontal Res.* 43(5): 531–536.

d'Aquino R, De Rosa A, Lanza V, Tirino V, Laino L, Graziano A, Desiderio V, Laino G, Papaccio G (2009). Human mandible bone defect repair by the grafting of dental pulp stem/progenitor cells and collagen sponge biocomplexes. *Eur. Cell Mater.* 18: 75–83.

Davies LC, Locke M, Webb RD, Roberts JT, Langley M, Thomas DW, Archer CW, Stephens P (2010). A multipotent neural crest-derived progenitor cell population is resident within the oral mucosa lamina propria. *Stem Cells Dev.* 19(6): 819–830.

Dupin E, Calloni GW, Le Douarin NM (2009). The cephalic neural crest of amniote vertebrates is composed of a large majority of precursors endowed with neural, melanocytic, chondrogenic and osteogenic potentialities. *Cell Cycle* 9(2): 238–249.

Dupin E, Glavieux C, Vaigot P, Le Douarin NM (2000). Endothelin 3 induces the reversion of melanocytes to glia through a neural crest-derived glial-melanocytic progenitor. *Proc. Natl. Acad. Sci.* USA 97(14): 7882–7887.

Dupin E, Real C, Glavieux-Pardanaud C, Vaigot P, Le Douarin NM (2003). Reversal of developmental restrictions in neural crest lineages: transition from Schwann cells to glial-melanocytic precursors *in vitro*. *Proc. Natl. Acad. Sci.* USA 100(9): 5229–5233.

Hu YF, Gourab K, Wells C, Clewes O, Schmit BD, Sieber-Blum M (2010). Epidermal neural crest stem cell (EPI-NCSC)-mediated recovery of sensory function in a mouse model of spinal cord injury. *Stem Cell Rev.* 6(2): 186–198.

Jinno H, Morozova O, Jones KL, Biernaskie JA, Paris M, Hosokawa R, Rudnicki MA, Chai Y, Rossi F, Marra MA, Miller FD (2010). Convergent genesis of an adult neural crest-like dermal stem cell from distinct developmental origins. *Stem Cells* 28(11): 2027–2040.

Kahnberg KE, Thilander H (1982). Healing of experimental excisional wounds in the rat palate. (I) Histological study of the interphase in wound healing after sharp dissection. *Int. J. Oral Surg.* 11(1): 44–51.

Kim JB, Greber B, Arauzo-Bravo MJ, Meyer J, Park KI, Zaehres H, Schöler HR (2009). Direct reprogramming of human neural stem cells by OCT4. *Nature* 461(7264) 649–643.

Kim JB, Sebastiano V, Wu G, Arauzo-Bravo MJ, Sasse P, Gentile L, Ko K, Ruau D, Ehrich M, van den Boom D, Meyer J, Hubner K, Bernemann C, Ortmeier C, Zenke M, Fleischmann BK, Zaehres H, Schöler HR (2009). Oct4-induced pluripotency in adult neural stem cells. *Cell* 136(3): 411–419.

King-Robson J (2011). Encouraging regeneration in the central nervous system: Is there a role for olfactory ensheathing cells? *Neurosci. Res.* in press.

Kondo T, Raff M (2000).Oligodendrocyte precursor cells reprogrammed to become multipotential CNS stem cells. *Science* 289(5485): 1754–1757.

Lavoie JF, Biernaskie JA, Chen Y, Bagli D, Alman B, Kaplan DR, Miller FD (2009). Skin-derived precursors differentiate into skeletogenic cell types and contribute to bone repair. *Stem Cells Dev.* 18(6): 893–906.

Le Douarin NM, Calloni GW, Dupin E (2008). The stem cells of the neural crest. *Cell Cycle* 7(8): 1013–1019.

Le Douarin NM, Teillet MA (1974). Experimental analysis of the migration and differentiation of neuroblasts of the autonomic nervous system and of neurectodermal mesenchymal derivatives, using a biological cell marking technique. *Dev. Biol.* 41(1): 162–184.

Lendahl U, Zimmerman LB, McKay RD (1990). CNS stem cells express a new class of intermediate filament protein. *Cell* 60(4): 585–595.

Li Y, Field PM, Raisman G (1997). Repair of adult rat corticospinal tract by transplants of olfactory ensheathing cells. *Science* 277(5334): 2000–2002.

Mackay-Sim A, Feron F, Cochrane J, Bassingthwaighte L, Bayliss C, Davies W, Fronek P, Gray C, Kerr G, Licina P, Nowitzke A, Perry C, Silburn PA, Urquhart S, Geraghty T (2008). Autologous olfactory ensheathing cell transplantation in human paraplegia: A 3-year clinical trial. *Brain* 131(Pt 9): 2376–2386.

Marynka-Kalmani K, Treves S, Yafee M, Rachima H, Gafni Y, Cohen MA, Pitaru S (2010). The lamina propria of adult human oral mucosa harbors a novel stem cell population. *Stem Cells* 28(5): 984–995.

Morrison SJ, Shah NM, Anderson DJ (1997). Regulatory mechanisms in stem cell biology. *Cell* 88(3): 287–298.

Murrell W, Wetzig A, Donnellan M, Feron F, Burne T, Meedeniya A, Kesby J, Bianco J, Perry C, Silburn P, Mackay-Sim A (2008). Olfactory mucosa is a potential source for autologous stem cell therapy for Parkinson's disease. *Stem Cells* 26(8): 2183–2192.

Nagoshi N, Shibata S, Kubota Y, Nakamura M, Nagai Y, Satoh E, Morikawa S, Okada Y, Mabuchi Y, Katoh H, Okada S, Fukuda K, Suda T, Matsuzaki Y, Toyama Y, Okano H (2008). Ontogeny and multipotency of neural crest-derived stem cells in mouse bone marrow, dorsal root ganglia, and whisker pad. *Cell Stem Cell* 2(4): 392–403.

Park D, Xiang AP, Mao FF, Zhang L, Di CG, Liu XM, Shao Y, Ma BF, Lee JH, Ha KS, Walton N, Lahn BT (2010). Nestin is required for the proper self-renewal of neural stem cells. *Stem Cells* 28(12): 2162–2171.

Pellegrini G, Rama P, Mavilio F, De Luca M (2009). Epithelial stem cells in corneal regeneration and epidermal gene therapy. *J. Pathol.* 217(2): 217–228.

Raff M (2003). Adult stem cell plasticity: Fact or artifact? Annu. Rev. *Cell Dev. Biol.* 19: 1–22.

Shakhova O, Sommer L (2010). *Neural crest-derived stem cells.* May 4, StemBook, ed. The Stem Cell Research Community. doi/10.3824/stembook.1.51.1.

Sieber-Blum M, Cohen AM (1980). Clonal analysis of quail neural crest cells: They are pluripotent and differentiate *in vitro* in the absence of noncrest cells. *Dev. Biol.* 80(1): 96–106.

Sieber-Blum M, Grim M, Hu YF, Szeder V (2004). Pluripotent neural crest stem cells in the adult hair follicle. *Dev. Dyn.* 231(2): 258–269.

Sieber-Blum M, Schnell L, Grim M, Hu YF, Schneider R, Schwab ME (2006). Characterization of epidermal neural crest stem cell (EPI-NCSC) grafts in the lesioned spinal cord. Mol. *Cell Neurosci.* 32(1–2): 67–81.

Sieber-Blum M (2010). Epidermal neural crest stem cells and their use in mouse models of spinal cord injury. *Brain Res. Bull.* 83(5): 189–193.

Stemple DL, Anderson DJ (1992). Isolation of a stem cell for neurons and glia from the mammalian neural crest. *Cell* 71(6): 973–985.

Takahashi K, Yamanaka S (2006). Induction of pluripotent stem cells from mouse embryonic and adult fibroblast cultures by defined factors. *Cell* 126(4): 663–676.

Trentin A, Glavieux-Pardanaud C, Le Douarin NM, Dupin E (2004). Self-renewal capacity is a widespread property of various types of neural crest precursor cells. *Proc. Natl. Acad. Sci.* US A, 101(13): 4495–4500.

Ueda M, Yamada Y, Kagami H. Hibi H (2008). Injectable bone applied for ridge augmentation and dental implant placement: Human progress study. *Implant Dent.* 17(1): 82–90.

Widera D, Grimm WD, Moebius JM, Mikenberg I, Piechaczek C, Gassmann G, Wolff NA, Thevenod F, Kaltschmidt C, Kaltschmidt B (2007). Highly efficient neural differentiation of human somatic stem cells, isolated by minimally invasive periodontal surgery. *Stem Cells Dev.* 16(3): 447–460.

Widera D, Mikenberg I, Elvers M, Kaltschmidt C, Kaltschmidt B (2006). Tumor necrosis factor alpha triggers proliferation of adult neural stem cells via IKK/NF-kappaB signalling. BMC *Neurosci.* 7: 64.

Widera D, Zander C, Heidbreder M, Kasperek Y, Noll T, Seitz O, Saldamli B, Sudhoff H, Sader R, Kaltschmidt C, Kaltschmidt B (2009). Adult Palatum as a Novel Source of Neural-Crest Related Stem Cells. *Stem Cells* 27(8): 1899–1910.

Wilkie AO, Morriss-Kay GM (2001). Genetics of craniofacial development and malformation. *Nat. Rev. Genet.* 2(6): 458–468.

Wu JP, Kuo JS, Liu YL, Tzeng SF (2000). Tumor necrosis factor-alpha modulates the proliferation of neural progenitors in the subventricular/ventricular zone of adult rat brain. *Neurosci. Lett.* 292(3): 203–206.

INDEX